Humanity's Footprint

Humanity's Footprint

Momentum, Impact, and Our Global Environment

Walter K. Dodds

 Columbia University Press   New York

Columbia University Press
Publishers Since 1893
New York Chichester, West Sussex

Copyright © 2008 Columbia University Press
All rights reserved

Library of Congress Cataloging-in-Publication Data
Dodds, Walter Kennedy, 1958–
Humanity's footprint : momentum, impact, and
our global environment / Walter K. Dodds.
p. cm.
Includes bibliographical references and index.
ISBN 978-0-231-13966-3 (cloth : alk. paper)
ISBN 978-0-231-13967-0 (pbk. : alk. paper)
ISBN 978-0-231-51304-3 (ebook)
1. Nature—Effect of human beings on. 2. Environmental degradation.
3. Overpopulation. 4. Environmental ethics. 5. Environmental justice.
6. Sustainable development. I. Title.

GF75.D65 2008
304.2—dc22
2007029338

∞

Columbia University Press books are printed on permanent and durable acid-free
paper.
Printed in the United States of America
c 10 9 8 7 6 5 4 3 2
p 10 9 8 7 6 5 4 3 2

For my family

Contents

Preface   ix

Acknowledgments   xiii

ONE
Collision Course: An Expanding Appetite for Resources Coupled
with Population Growth on a Finite Planet   1

TWO
The Insidious Explosion: Global Trends in Population Growth
and Resource Use Expand Our Footprint   11

THREE
Shock Wave from the Insidious Explosion:
Momentum of Human Effects on Our Planet   34

FOUR
Weeds and Shrinking Violets:
Pests on the Move and the Ecological Holocaust   63

FIVE
Survival on a Finite Earth: The Ultimate Game,
or Why Human Nature Destines Us to Use More Than Our Share   83

SIX
Why Humans Foul the Nest:
Cultural and Genetic Roots Run Deep   108

SEVEN
Searching for Answers:
Can We Achieve Sustainability, or Are We Screwed?   136

EIGHT
No More Business as Usual:
Transcendence, Enlightenment, Rationalization, Hope, and Action   171

NINE
Consilience: Socioenvironmental Restoration
and Sustainable Inhabitation of Earth   189

APPENDIX 1
Data Sources Used to Make Graphs   215

APPENDIX 2
Reading the Graphs in This Book   221

APPENDIX 3
Putting Global Environmental Impact into a
Quantitative Framework That Links Impact and Behavior   225

APPENDIX 4
Some Organizations Involved with
Global Environmental Issues   229

Notes   235

Index   263

Preface

This book can be viewed as the most general of self-help books: a look at the health of the Earth and how we can maintain (even restore) that health. An insidious explosion of human population growth and resource use is creating a shock wave that is reverberating through our global environment for the first time in the history of humanity. The cumulative effect of these influences, humanity's footprint, has led us to overshoot the capacity of Earth to support us. Unfortunately, we are lost in a sea of issues, each of which needs immediate attention. For at least one in every six people on our planet, the next meal is the primary concern. For those fortunate enough to have reliable food and shelter, providing stability for the future and some chance that all people will have a humane standard of living assumes greater importance. For most people in developed countries, security means achieving economic well-being and controlling perceived danger from threats such as disease, crime, and terrorism. To be sure, these threats are high profile, generating intense media coverage and pushing other issues to back burners, but the world's ultimate problems are based on factors far larger in scale than most of those issues that are highlighted in the daily news. If we are to ensure security for future generations, our society must find a way to deal with the fact that our actions have consequences on a global scale. The approach must be to weave environmental knowledge with information from other fields (for example, psychology, anthropology, and economics) to synthesize the causes of complex problems and to suggest potential solutions.

This book is my attempt to highlight the most fundamental aspects of modern society related to global environmental issues. With my training, facts form the basis for any critical analysis of a problem. As a scientist, I present information in this book as accurately as possible. Scientists also know that a degree of uncertainty is associated with all knowledge. Thus the book is generally arranged with the information that I think is the most reliable at the beginning, with more speculative ideas at the end.

I also try to present an unbiased view of global environmental problems and solutions. Scientists are acutely aware of personal bias. Facts and ideas in this book are presented as objectively as possible. David Pimentel, a professor of entomology at Cornell University, told me that he was not concerned about environmental issues early in his career. He attended a seminar where the speaker described environmental threats, and he thought that the speaker must be biased. At the time, he thought that if even a portion of what the speaker said was true, environment and humanity were potentially in real trouble. He checked facts in the scientific literature and realized that our environment was in even worse shape than the speaker had indicated. This led Dr. Pimentel to a distinguished career as one of the most internationally respected scientific environmental watchdogs. So beware of biases, and check my facts if you doubt them. Humanity could be headed for real trouble if only some of the threats that I describe are true. Just like fire alarms, even if some are false, it is not a good policy to ignore them if you want to protect a community against catastrophic fires. By suggesting that readers check my facts, I hope to avoid the appearance of "crying wolf." For some people, this book could serve as the fire alarm that opens their eyes to the potential challenges for humanity with regard to the global environment.

I employ several guidelines to logically and critically assess data in this book. The following three principles helped me avoid bias and may help the reader as well:

1. The more certain someone is about the future and the more long term their forecasts, the less likely they are to be correct.[1]

2. Just because two things are correlated does not mean that one causes the other, but considering such correlations is the first step toward understanding our world.

3. The more people have to gain from distorting the truth, the more likely they are to do so. (For example, you are more likely to believe you need a specific drug when it is suggested by a doctor than by a representative of a pharmaceutical company that profits from selling the drug.) Truth should not be governed by someone's vested interest.

In this book, I attempt to present a simple, understandable account of global environmental trends, the immediate and ultimate reasons why they are occurring, and our prospects for achieving a sustainable world with a lifestyle equivalent to what those in developed countries currently enjoy. Although the tone may strike some as overly pessimistic, I paint as realistic a picture as possible of potential global environmental problems and their causes, and I present the challenges humans must face in solving them. The label "doom and gloom environmentalist" has become a convenient way to dismiss a bearer of bad news, regardless of how true the news is. Simply identifying problems is only a small part of solving problems. Thus, I also suggest how humans can deal with problems on a global scale, and I provide reason to hope we will be able to control our global environmental footprint.

Acknowledgments

I thank the many people who helped me form and integrate the ideas presented, both those who had discussions with me and those whose writing has provided information and inspiration. The members of the Kansas State University Aquatic Journal Club particularly spurred my thinking, while the Little Apple Brewery provided affordable and delicious libations to stimulate discussions. David Pimentel showed me that it is possible to do something about problems, provided helpful reviews, and pointed out that the population explosion is insidious. Thanks to Melody Bernot, Randy Bernot, Lois Domek, Nicole Gerlanc, Mike Quist, and Chris Smith for closely reading an early version and making helpful suggestions. Becky Clark and Rachel Sherck test-drove some of the chapters, and John Blair helped with appendix 3. Many thanks to Andrew Blaustein, Robert Costanza, Roger H. von Haefen, James Karr, David Orr, and David Pimentel for careful and helpful reviews. Several anonymous reviewers also made very useful suggestions, as did Chuck Crumly and Ron Laitch. Patrick Fitzgerald contributed many great ideas and guided me through this project. Any shortcomings in the book are mine. Thanks to the team at Columbia University Press (Milenda Lee, Irene Pavitt, and Cynthia Garver) for moving this project to completion and improving it along the way. Thank you to

Josiah Boggs Dodds for two readings and many helpful ideas and comments on the text, and for shaping who I am as a person. Thanks to Steve Carpenter for pointing out the key points of the Millennium Ecosystem Assessment. Don Kaufman provided the starship-without-nature example. Some of the material in chapter 5 appeared previously in "The Commons, Game Theory, and Aspects of Human Nature That May Allow Conservation of Global Resources," *Environmental Values* 14 (2005): 411–426, and I thank the journal for graciously allowing its publication here. I am grateful to the Division of Biology at Kansas State University and the Kansas Agricultural Experiment Station for providing my salary during the years this book was being written. This is publication 06-264-B from the Kansas Agricultural Experiment Station. Thanks to my children, whose future I was thinking about while writing this book, and to my wife, Dolly, who read multiple drafts, made the whole thing better, and is always there for me.

Humanity's Footprint

1

Collision Course

An Expanding Appetite for Resources Coupled with Population Growth on a Finite Planet

Power, wealth, and security are inextricably linked to what our Earth provides us and how we treat the planet that sustains us. In the sixteenth century, power and wealth revolved around trade, with the nations of Europe (the global superpowers of the day) racing to establish reliable and quick trade routes with India, China, and Japan. The search for the Northwest Passage to circumvent South America and sail directly to Asia began in the early sixteenth century and continued until the beginning of the twentieth, when, in 1906, Roald Amundsen, the famous Norwegian explorer, sailed a dangerous, iceberg-cluttered route from Norway to what is now Alaska. But that route was too hazardous and shallow to be used as a trade route, and it was not until the 1950s that a medium-size icebreaker navigated the Arctic Ocean north of Canada. In the past few years, our climate has warmed so much that now, during summer, the Arctic ice sheet melts and commercial vessels are finally able to traverse the Northwest Passage (figure 1.1). By the summer of 2007, the ice sheet had melted to about half the size it was 25 years earlier.

We can only imagine how history would have changed if the climate had been warm enough in the sixteenth century to allow free passage north from North America to India. Would Canada be more wealthy and powerful than the United States? Would Britain have focused even

1979

2005

FIGURE 1.1 The Arctic ice cap in (*top*) summer 1979 and (*bottom*) 2005. The ice cap along the lower left side (above Alaska) has retreated about 400 miles (650 km). (Images courtesy of NASA)

more attention on Canada? Would the French have been willing to offer the Louisiana Purchase? Today's questions are similar: Given current trends, will polar bears go extinct if polar ice sheets continue to diminish? Will native Inuit (Eskimos) or other Arctic indigenous people maintain any semblance of their traditional ways? Will the balance of power among Earth's nations shift in response to global climate change? Global climate has clear and strong effects on human affairs and the natural world.

A fire from the burning remains of slashed rainforest so large it can be seen from space, sheets of ice the size of small countries breaking off Antarctica, fish in remote Arctic lakes too contaminated with toxic chemicals to be safely eaten, and precipitous declines of every major ocean fishery—these are just a few signs of the pervasive influence of humans on our world. Plastic waste from humans can be found in every ocean and on every beach on Earth.[1] Global impacts of humans on their environment, such as the greenhouse effect and increases in ultraviolet radiation caused by ozone depletion, are reality.

I define "environment" in this book as physical, chemical, and biotic features that affect organisms, including humans, and determine their survival and the resources available to them. Humanity's footprint (the sum of our environmental impact) has likely surpassed the capacity of Earth's environment to sustain our actions, and our global impacts continue to increase.[2]

Bill McKibben, who has written extensively about environmental issues, has referred to pervasive human impacts as "the end of nature."[3] By this he means that wild nature, unaffected by man, no longer exists. Human influences extend to the most remote regions on Earth. Pesticides are found in animals in the open ocean and far north in the Arctic, areas far from human habitation. As we continue to alter global climate, no ecosystem will remain uninfluenced. Changes will be profound, long lasting, and pervasive.

Humans are moving forward into a global environmental future with tremendous momentum. Momentum is what makes a runaway train so dangerous; even immediate application of the brakes does not stop the train for many miles. Our environmental momentum may have already overshot the capacity of Earth to support us. This environmental momentum occurs as a result of the magnitude of global processes and because humans are accelerating numerous global trends. Many trends, fundamentally driven by increases in global population and the amount and efficiency of resource use, are influenced by the enormity of this momentum. I try in this book to define the trajectory of the missile fired by the environmental catapult (where humanity is going), the momentum of our global environmental impact, and our ability to control the trajectory in the face of the tremendous momentum.

More than 6.5 billion people currently live on Earth. People use the energy equivalent of more than 33 billion U.S. tons (30 billion metric tons) of oil each year. The global economy (annual global domestic product) has reached more than \$25 trillion. How many is 6.5 billion people, how much is 33 billion tons of oil, and how much is \$25 trillion? If a particular special person is one in a million, there are currently over 6500 special people just like them in the world. Picture a stadium that holds 50,000 people. Two stadiums would hold 100,000 people; 20 would hold 1 million; 20,000 would hold 1 billion; and 130,000 would be needed to hold all the people on Earth. I have difficulty conceiving of 130,000 stadiums, let alone 130,000 stadiums with 50,000 people in each.

Assuming that a large stadium can hold a volume of 110 million cubic feet (3.1 million m^3) of liquid, the approximate capacity of the Cleveland Browns' football stadium, the volume of oil required to provide the equivalent of energy used by people on Earth each year would fill more than 11,000 such stadiums. Stacking 1 million \$1 bills, each 0.0043 inch (0.1 mm) thick, would result in a stack 358 feet (109 m) tall. A stack of 25 trillion bills—the gross global product—would be 1.7 million miles (2.7 million km) high, 3.5 round trips from Earth to the moon. Picture 10,000 dots (figure 1.2). If six of these groupings (60,000) fill a page, 300 million dots—the current population of the United States—would fill a book of 500 pages, and the human population on Earth would be an encyclopedia over 10,000 pages long. It takes about 20 seconds to count 100 dots; it would take 41 years to count 6.5 billion dots, assuming no breaks for meals or sleep. Our brains are poorly equipped to deal with such large numbers.

Although knowledge of our global effects is in the public consciousness, it is at a superficial level. Scientists have just begun documenting global environmental problems, their causes, some of their implications, and what it may take to reverse them. Global environmental effects of human behavior have been studied for only the past several decades, so that the implications of these global effects for the future of humanity are not yet known.

Peter Vitousek, a professor at Stanford University, was named among America's best scientists by *Time* and CNN in 2001. He and his colleagues published several articles documenting humanity's global influences on how Earth's ecosystems function.[4] They found that we humans directly or indirectly use half of all products of photosynthesis on Earth, we now cause nitrogen and phosphorus to enter global biological systems at twice the natural rates, and we are altering our atmosphere for decades to come. Some of these papers have been cited more than 500 times in the scientific literature over the past ten years, which

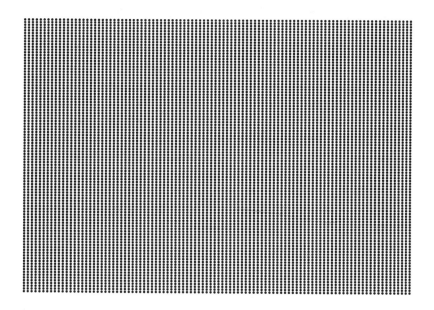

FIGURE 1.2 Image of 10,000 dots. If 6 of these fill a page, how long will a book with 6.5 billion be?

are very high rates of citation for scholarly works. The attention these papers received, and the research they stimulated, illustrate that environmental scientists have only recently started to pay close attention to global environmental impacts.

A recent massive effort to account for human effects on the global biosphere, the Millennium Ecosystem Assessment, documents the human influences on the Earth that supports us.[5] This effort involved a global assessment by 1360 authors from 95 countries. The assessment was reviewed by an independent review board of 80 experts guiding 850 reviewers. The group found that 15 of the 24 ways that Earth supports humanity are degraded: fisheries, wild foods, fuel wood, genetic resources, biochemicals, medicines, water supply, air quality, regional climate, erosion, water purification, pest regulation, pollination, natural hazard regulation, and spiritual, religious, and aesthetic values. We are entering the "overshoot century," where we are exceeding the capacity of Earth to support humanity.

Over the past 300 years, humans have caused unprecedented changes in and on the land, oceans, and atmosphere. For the first time in history, humans are having multiple effects on the global environment that sustains us, and these effects are accelerating. Human behaviors causing these effects are deeply rooted and will be extraordinarily difficult to change. Past societies have destroyed their local environments, but members of our species, *Homo sapiens*, are having effects that far exceed those documented at any other time in human history. We are altering the entire Earth that supports our lives and doing so without complete understanding of the ultimate effects of such alteration. Close your eyes, step off the edge, and see what happens. Is it a low curb or a gaping chasm? Are we hurtling toward collision with a nearby brick wall, or will our momentum and trajectory end in a soft landing far in the future? Humanity is just reaching the point where we can begin to understand the repercussions of our actions. Understanding may allow realistic hope for the future.

What Are the Biggest Problems on Earth and Their Root Causes?

A popular question when interviewing young people for achievement awards is: What are the biggest problems on Earth, and what will it take to solve these problems? This question implies a more personal question to the interviewee: What are you going to do to solve the world's greatest problems? If all people seriously ask themselves this question, the world might end up a better place. After I had asked many students how they were going to solve Earth's problems, it occurred to me that maybe I should ask myself the same question. This questioning led to thinking in the broadest and most general terms about the planet's ability to support us.

My personal opinion is that the largest problems facing people on Earth are poverty (including the ability to feed humanity), the threat of nuclear war, the catastrophic spread of diseases from bioterrorism or human behaviors that encourage emergent diseases, and the effect of human pressures on Earth's ability to sustain us. I do not know which of these problems is most important, but, clearly, poverty directly harms the most people right now. According to the World Health Organization in 2007, some 54 percent of the 10.8 million deaths per year of children under age five are directly or indirectly caused by severe malnutrition. Over 5 million children die each year, even though we have food to feed them. I do not have the expertise to offer insights into global poverty or the political nuance that fails to control hunger and that, so far, has controlled the danger of annihilation by biological or nuclear warfare and by epidemics of human diseases. As a biologist who studies the environment, I can most easily approach global environmental problems. This exploration took me into human behavior, particularly as influenced by biology.

"Save the Earth" is the slogan that most succinctly expresses environmental concern over global effects of humans, but does Earth truly need saving? As a scientist who studies freshwater and the plants and animals that live in it, I am astounded by the immense cumulative increase in evidence of human effects on our environment. Specialists are individually documenting negative impacts of humans on every ecological system on the planet. On land and in the sea, in remote areas and near large cities, on mountaintops and deep below Earth's surface, the effects of human activities are becoming ever more pronounced. Each scientific meeting I attend and each research journal I read offers further proof of the collective negative effects of humans on our globe. Every human inhabitant on Earth has only a small impact, but the sum of 6.5 billion impacts is taking a serious toll on our world.

Appreciating trends that are occurring on a global scale is difficult because they are abstract, involve large geographic areas, require accounting with gigantic numbers, and are generally far removed from our everyday experience. Although basic understanding of the science of ecology is not strong among the general public, good introductory, nontechnical treatments exist.[6] In addition, people living in North America and western Europe may not realize the degree of global degradation because in their countries many local environmental impacts have been minimized. A view biased by living in well-off countries allows technological optimists to claim that an increased standard of living leads to a healthier environment. The healthier environment myth evaporates when global accounting is used, but many traditional economists are partial to a worldview that discounts environmental costs. Negative costs of economic activities (called "externalities") are difficult to assess. If you ignore externalities, and money is flowing into your pocket, things can look pretty good—in fact, much better than they actually are.

Unfortunately, environmental conditions are appalling in most developing countries because pollution controls are often not enacted or are poorly enforced in the face of poverty.[7] Although the distinction between developed and developing countries is simplistic, in this book, I use the term "developed countries" to mean the United States and Canada and those in western Europe, along with Japan, Australia, and other countries with a relatively high standard of living. The least-developed or developing countries are in Africa, Asia (excluding Japan),

Latin America, the Caribbean, and the regions of Melanesia, Micronesia, and Polynesia. Two-thirds of the people on Earth live in developing countries, and about one-half of those live in extremely crowded and polluted cities.

People everywhere care about their environment. Gallup International surveyed 50,000 people from 60 countries for its Millennium Survey and found that health and family life were most important to respondents. A majority, 57 percent, think the current state of the environment is unsatisfactory. People in only five of 60 countries think their government is doing enough to protect the environment.[8] Why do people say they care about the environment but continue to damage it and, potentially, their own health? In part, the disconnect between environmental concern and actual behavior could stem from the inability to comprehend environmental trends at such a large scale and the true implications of those trends.

The ultimate reasons for human-caused environmental damage on a global scale are the patterns and ecological impact of resource use and population growth. Resource use must be considered because it is not simply the number of people on Earth but the effect of each person that determines environmental momentum.[9] Trends in human population growth and resource use at the global scale have a tremendous amount of momentum because of the sheer number of people involved and the size of the planetary ecosystem. The longer we wait to control the root causes of global environmental problems, the more momentum they gain and the less likely we are to be successful in protecting the Earth that supports us. Is it too late? If I thought so, there would be little point in writing this book.

Optimism, Pessimism, and Hope

How do we respond to environmental issues that are so large? Many people react strongly against "gloom and doom" environmentalists. Complete pessimism leaves no room for hope. The dichotomy in the environmental debate over global human effects has been well defined since the 1960s, when concern over the rapidly expanding human population led to the popular concept of the "population bomb" and the idea that we are reaching the limits of Earth to support us.[10] Alternatively, some technological optimists argue that science will solve any problems and that practical limits on human population and its rate of resource use do not exist. Perhaps optimists are correct, but prudence suggests that we should at least investigate our potential global effects and control the behaviors that could cause long-term damage to the ecosystems that support us. As world expert on biodiversity Peter Raven, director of the Missouri Botanical Garden, says, "Being optimistic about the future by wearing rose-colored glasses and engaging in wishful thinking in a moral vacuum constitutes a crime against our posterity."[11]

As in most public debates, reality probably lies between the position of extreme environmental alarmists (who claim we are certainly doomed if we do not solve all major environmental problems immediately) and environmental deniers (who deny the existence of any major threats to humanity from environmental problems). Both the alarmists' and the deniers' voices are loud. They compete for the limelight in this world, where media vies for our attention every minute. Regardless of political rhetoric, strong scientific documentation of global environmental threats exists, and those threats are described in this book. As a prerequisite to realistic hope, understanding (skeptical optimism) is logically better than blind faith.

I am optimistic that humans have the capacity to mitigate the negative effects of our global impacts but somewhat skeptical that we have the collective will to do so. I am optimistic that the human race will survive for many more generations but skeptical that the quality of life will be sufficient to allow general happiness. I am optimistic because people seem to be becoming aware of our global environmental impacts but skeptical that the rich and powerful have anything to gain from moderating those impacts.

As with any political or societal issue, the immediate effects of global environmental problems must be understood, the root causes of problems firmly grasped, and the solutions formulated based on scientific realities and societal constraints. In this book, I document the global environmental impacts caused by humans, propose aspects of basic human nature that create such impacts, and explore the possibilities for attaining sustainable use of the planet we inhabit. I substantiate the global environmental changes that are occurring on a large scale, discuss how the problems will take from decades to centuries to reverse, and explore how our current circumstances are related to deeply ingrained human behaviors. This book is intended for the layperson who is thinking about what he or she can do to improve the lot of humanity. Now that we live in a global society, it is increasingly imperative that our species views its problems and solves them from a global perspective. Technology has provided some of the tools to help us comprehend global trends. As humans, we can use our intellectual capabilities to overcome our limited capacities to sense the world around us and the changes we are making.

Environmental impacts ultimately are driven by increases in damaging patterns of resource use and rapid continued population growth, as discussed in chapter 2. In chapter 3, I describe some of the cumulative physical and chemical environmental impacts that humans have on Earth and how they correlate with trends of resource use rates and human population growth. In chapter 4, I document how humans cause the extinction of many species and introduce undesirable species at the same time. In chapters 5 and 6, I explore how basic human behaviors inevitably lead to environmental pressures, with particular emphasis on the numerous recent advances in the study of human behavior. Strategies for exploiting a limited resource base and how this relates to resource consumption form the basis of these chapters. In chapter 7, I suggest what can be done to use Earth's resources sustainably, and in chapters 8 and 9, I examine the possibilities for humans to manage their impacts, then provide suggestions for how to protect our Earth.

The issues revolve around a conflict between two central ideas related to Earth's capacity to support us:

1. The best predictor of future behavior of large groups of people is prior behavior (cultural trends have inertia in part because they are driven by repeatability of human behavior).
2. Earth has finite resources and fragilely balanced ecosystems.

Thus deeply rooted patterns of behavior lead to population growth and increased resource use. Humanity's footprint exceeds the Earth's capacity for support. These behaviors put us on a collision course that we cannot continue indefinitely in a finite world.

When families have spent a million years making nine
in the hope of saving one, they cannot stop making nine.
Culture is a slingshot moved by the force of its past. When
the strap lets go, what flies forward will not be family
planning, it will be the small, hard head of a child.

—Barbara Kingsolver

2

The Insidious Explosion

Global Trends in Population Growth and
Resource Use Expand Our Footprint

Growth in human population and resource use rates is undeniable. In
this chapter, I document those rates and the global momentum behind
increases in standard of living and population growth. Although the
numbers for population growth are solid and available, the rates of resource use, habitat destruction, and other human activities contribute
as much to the cumulative impact of humans on Earth. But because it
is difficult to account for their exact impacts on the environment, resource use rates are emphasized here. For example, calculating how
much environmental damage each gallon of oil burned will cause depends on numerous factors, such as how it is burned (Did the car have
emission controls?) and what it was used for (Was it used to run a
pump cleaning up an oil spill?). One method that attempts to provide
such a calculation is the determination of an ecological footprint for
each individual and for the global human population.

The Population "Bomb" Is Still Exploding

Population growth has drastically increased, leading to the idea of a
population bomb, which gained popularity in the 1960s. This ongoing

and insidious explosion is evident only when viewed over decades and across the globe, however, leading some to deny that explosive growth causes any problems. From an individual perspective, the problem is barely perceptible because we experience the effects of population growth so remotely. For example, even though older people remember a time when it seemed as if there were far fewer people, many may have seen only a small part of the planet. In fact, population growth is a massive, inexorable, and insidious trend. But because of scale, we can be part of an "explosion" and not be able to perceive it. To be sure, we are living in the biggest ongoing explosion of all, the Big Bang in an expanding universe, but humanity remained oblivious of its existence until we had the tools to observe it. Likewise, we now have tools to assess global trends in human population, resource use, and environmental impacts. These tools allow me to report on how humans have spread and dominated the planet over the past few centuries.

Human population is increasing to unprecedented levels (figure 2.1; see also appendix 1 and, for those who need general help in interpreting graphs, appendix 2), and this trend will probably continue through the lifetimes of all but the youngest readers of this book. In the nineteenth century, there were fewer than 2 billion people. In March 2006, the U.S. Census Bureau estimated the world population as more than 6.5 billion people. Increase in human population is the most obvious and well-documented global trend related to human effects on Earth. Linked hand to hand, 6.5 billion people would circle Earth's equator roughly 197 times (assuming an ability to stand on water). Each day, approximately 155,000 people die, and 360,000 people are born.[1] This translates to 108 deaths and 244 births per minute, 1.8 deaths and 4.1 births per second, or about 10 deaths and 25 births in the time it took you to read this sentence. The surplus of births over deaths is why the population is increasing by more than 200,000 people per day. Growth continues as it has for centuries.

Growth rate is geometric (sometimes referred to as exponential). Between 1900 and 2000, the actual growth rate accelerated (to become superexponential) as human health improved and mortality decreased. The explosive population growth illustrated in figure 2.1 led Paul Ehrlich to write his 1968 book *The Population Bomb*, which described some potential outcomes of that growth. Since then, Ehrlich has been a strong proponent for environmental responsibility and preservation while continuing an internationally acclaimed academic research career on insect ecology and evolution. Professor and current head of the

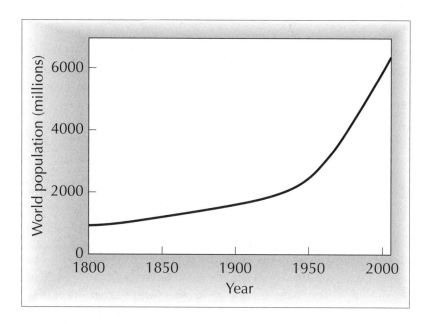

FIGURE 2.1 Human population from 1800 to 2005, showing the approximate 2.5-fold increase since 1950. (For data sources in this and other similar graphs without cited sources, see appendix 1, table 1.)

Center for Conservation Biology at Stanford University, he is a respected scientist and has received numerous environmental and research awards. Although a successful best seller, *The Population Bomb* was controversial, and some of Ehrlich's more catastrophic potential scenarios have not come to pass. For example, Ehrlich lost a famous bet with an economist, Julian Simon: Ehrlich bet that five metals would cost more in the future as we ran out of these resources, and he was wrong.

What we did not know at the time was that the human impact on the global environment would manifest itself as global warming, ozone depletion, and water scarcity, among other things. The fact remains that unlimited growth is not possible on a finite planet; the population has almost doubled since Ehrlich wrote his book, and we are experiencing global effects related to the greater population. Global population approximately doubled in the 150 years after 1800; then, in the 50 years from 1950 to 2000, it expanded more than threefold. It is irrefutable that this growth cannot continue indefinitely, as has been known for at least 200 years, when Thomas Malthus pointed out the logical limits to exponential growth.[2] What Malthus may have failed to note is the catastrophic failure of populations when they overshoot their limits.[3] The number of people Earth can actually support is a matter of contention.[4] Still, a conflict between the following two basic concepts remains:

1. The best predictor of future human behavior is past behavior.
2. Earth's resources are finite.

Humans cannot continue their past behavior of increasing numbers if they are restricted to a finite planet.

A simple example illustrates that current population growth cannot be maintained. Start with a pair of baby mice of opposite sex, and assume unlimited food; the animals can mature and produce four babies in three months, and all adults survive to reproductive age and die after reproduction. After three months, there will be four mice; in six months, eight mice; and so on. After two years, there will be 512 mice; after three years, approximately 8192 mice; and after four years, more than 131,000 mice. Assuming the mice are 1.5 inches long, 1 inch wide, 1 inch tall (5.1 by 2.5 by 2.5 cm), and box shaped (easier for the purposes of this calculation than assuming a spherical mouse), they would cover all land on Earth 1 inch (2.5 cm) deep by 12 years; after 15 years, they would be more than 680 feet (207 m) deep. This rate of growth obviously could not be supported indefinitely. Now assume one boxcar of food is supplied per day, and each mouse consumes one-millionth of a boxcar of food each day. At the beginning, mice will grow quickly. Once the population reaches about 1 million, food becomes limiting. At this point, the number of mice will be controlled by starvation, cannibalism, disease, and reproductive failure. Mice cannot grow forever at their maximum potential rate because something will control their population eventually (there is a conflict between the reproductive behavior of mice and the finite nature of the world they inhabit).

Likewise, no reasonable person thinks that human population growth on Earth can continue unabated. The question is not if we should limit our population, but how and when we should limit it. Either we will limit our total population by choice, or the capacity of Earth to support us will eventually constrain our population for us. Limitation by mechanisms we do not choose could be very unpleasant, including possible economic calamity, societal collapse, and vast human suffering. History has demonstrated that human suffering is a common result of environmental overexploitation, combined with other factors. This was likely true for the inhabitants of Easter Island, the Anasazi of the southwestern United States, and the Maya of Mexico and Central America; disastrous overexploitation of the environment is occurring now in Haiti and Rwanda.[5]

One very important aspect of limiting human population is the time lag between initiating a control and when a population actually stops increasing. This time lag in the face of control measures is referred to as the momentum of population growth. Momentum occurs because children born today will likely have children in the future, even if each child has fewer offspring than their parents did. Excess of births over deaths is a fundamental characteristic of a growing population, and it leads to a proportionally greater number of children. Therefore, even if all human couples abruptly start to have only enough children to replace themselves, most children who are not yet reproductive will reach puberty and have their children before Earth's population stabilizes.

China exemplifies the time lag inherent in limiting population growth. In 1968, each woman in China had an average of 6.5 children. By 1980, strict enforcement and massive education, as well as social pressure, led to each woman having an average of 2.2 children; by 1992, each woman was averaging 1.9 children.[6] The Chinese population was 830 million in 1970, 1.155 billion in 1990, and 1.275 billion in 2000.[7] Thus, in spite of reaching replacement fertility, the Chinese population has continued to grow for two decades and will continue to grow for at least the next 20 years. This is population momentum.

Slow transition to a stable population also applies to global population trends. If women had an average of 2.06 children from now on (the global average is currently 2.8), Earth would reach a stable population in about 100 years, assuming that advances in standard of living and health care will lead to slightly lower than current mortality rates.[8] Stable population would be reached eventually because, on average, about 0.06 of a person dies for every two who reach reproductive age. Until population stabilizes, however, we will have a global population with many children who will reproduce when they come of age. Thus population growth momentum will cause global population to reach at least 8 billion people (a one-third increase) before it stabilizes.[9]

An estimate of a stable population of 8 billion humans within the next 100 years assumes that we could somehow reach worldwide replacement fertility quickly, but nothing currently indicates that we will do that any time in the near future. Momentum of human population growth is catapulting us into the future, and population policy decisions made now will require decades or centuries to take effect on the global scale. So if we cannot reach replacement fertility on Earth right away, how large will the human population get, according to demographers? As with any prediction, numbers are uncertain. Catastrophic mortality from a global pandemic caused by a new disease could cut Earth's population substantially, as did the Black Death in the mid-fourteenth century. Certainly, our environmental impacts are increasing the probability of catastrophic pandemics, as discussed in chapter 4.

Several estimates of population growth, given no catastrophic epidemic, are available. Estimates made by the United Nations indicate that world population is now increasing by about 77 million people each year.[10] Six countries (India, 21 percent; China, 12 percent; Pakistan, 5 percent; Nigeria, 4 percent; Bangladesh, 4 percent; and Indonesia, 3 percent) account for one-half of this annual growth. By 2050, population is expected to be 9.3 billion people (low estimate, 7.9; high estimate, 10.9 billion people). Fertility in virtually all developed countries has fallen below replacement level (including countries in North America and western Europe, as well as Australia, Japan, and various other countries), but population growth will continue in developing countries. Although world average fertility rates have fallen from 5 to 2.8 children per woman, human population is not expected to reach replacement fertility until 2050. Thus Earth's population will continue to grow for at least 100 more years: 50 years to replacement level, and 50 more to reach stable population structure.

What level of human population can Earth ultimately support? This question is impossible to answer. Estimates vary widely, but the median value of a broad range of estimates of how many people Earth can support sustainably falls in the range of 7.7 to 12 billion.[11] Population *growth rates*, however, currently are decreasing. Figure 2.1 illustrates that the rate of population increase slowed only slightly over the past 10 years (that is, the curve did not rise quite as quickly as it did during the past 50 years). The slight decrease in growth rate does not mean that human population numbers are decreasing, just that the rate of population growth is not as great. Absolute population of humans is still growing by 211,000 people per day.

Rates of Resource Use Continue to Increase

In addition to human population, rates of resource use per person are increasing. "Resource use" is defined here as any good or service provided by Earth that humans use. This definition includes traditional uses (such as energy extraction and forestry) and encompasses use of Earth's capacity to sustain healthy human life, such as maintaining food production, as well as water and air quality. By this definition, forests are a resource not only for the trees they provide but also for the oxygen they produce through photosynthesis for us to breathe, the greenhouse gas emissions (carbon dioxide) they absorb, the useful organisms they harbor, and the water-quality benefits they confer. Forests are also a resource of less-tangible properties, such as recreational and cultural values.

Numerous strategies can be used to estimate resource use rate, each of which has its drawbacks and benefits. I use changes in gross economic activity, energy use, water use, and area of cropland as indicators of rates at which humans are using Earth's resources. All four indicators (and most others not covered here) portray a similar picture of resource use: *global rates of resource use are increasing rapidly.*

Hyperexpansion of the Economy

Global gross domestic product for each person per year (in U.S. dollars adjusted for inflation to 1990) has undergone a massive expansion over the past 200 years (figure 2.2). In the past 100 years alone, individuals have created fourfold more economic activity. Economic expansion is highly influenced by the wealthiest countries, but the standard of living has improved for many people living in lesser-developed countries as well. This trend of economic expansion over time is exponential. When per capita rate of increase is multiplied by number of people on Earth, we can calculate global economic productivity. Multiplying exponential increase in dollars per person by a superexponentially growing number of people results in a massive expansion in the amount of economic activity (more than sevenfold in the past 100 years, compared with a three to four times greater human population over the same period). Overall, economic activity has approximately doubled in just the past 25 years.

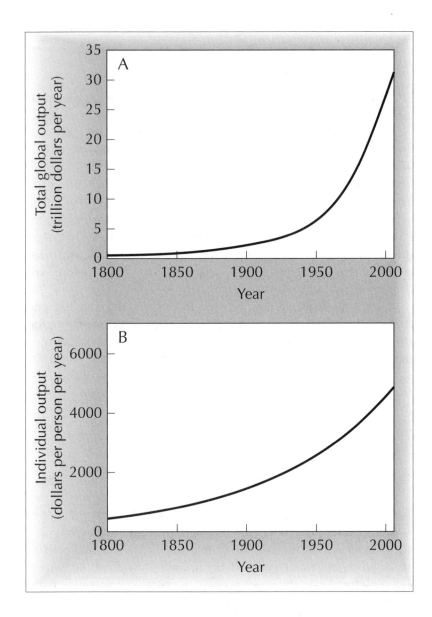

FIGURE 2.2 Economic output from 1800 to 2005: (*A*) total global output and (*B*) averaged per person, showing the huge economic expansion since 1950. Consider the environmental consequences that occur with this vast increase in standard of living.

Economic expansion has greatly increased the well-being of many humans on Earth. This astonishing growth in economic activity is tied to greater rates of resource use and has inevitable negative effects on the global environment. The relationship between global economic activity and other types of resource use must be delineated to estimate the impact associated with this impressive increase in economic activity.

Burgeoning Energy Consumption

The total rate of resource use can also be gauged by energy consumption. If all major forms of energy consumption are considered (most energy consumption is currently burning wood, coal, oil, and gas), increasing amounts are being used globally (figure 2.3). Energy use has risen approximately 600 percent in the past 100 years, close to the amount of growth in the global domestic product. Greater energy use is even more tightly tied to environmental effects than is expanded economic activity; energy use has resulted in all of the following: global lead emissions that are 15 times higher than natural, up to 10 times more oil spilled into oceans than enters through natural processes, a doubling in acid precipitation, global climate impacts (discussed in detail in chapter 3), and many other environmental impacts.[12] Just decreasing the amount of environmental impact per amount of energy used (cleaner energy use) still would leave a huge environmental impact. Both amount used and environmental impact per unit used should be decreased.

Average rate of energy use per person on Earth (see figure 2.3) presents a pattern similar to that for economic activity. Starting in 1800, average rates of energy use per person increased over what was possible when only wood fuel and limited machinery were available. As the industrial revolution took hold and mechanical means for production proliferated, rates of energy use escalated rapidly. First, more rapid rates of using wood were developed, then more coal was used, and finally oil and gas took over as primary sources of energy. Currently, about two-thirds of global energy is supplied by oil and gas. Some people on Earth still have energy use rates similar to those before 1800, but, in aggregate, rates have skyrocketed and are driven disproportionately by the inhabitants of developed countries.

As economic activity increases, so does the rate of energy consumption per person. Sport utility vehicles are capable of 200 to 300

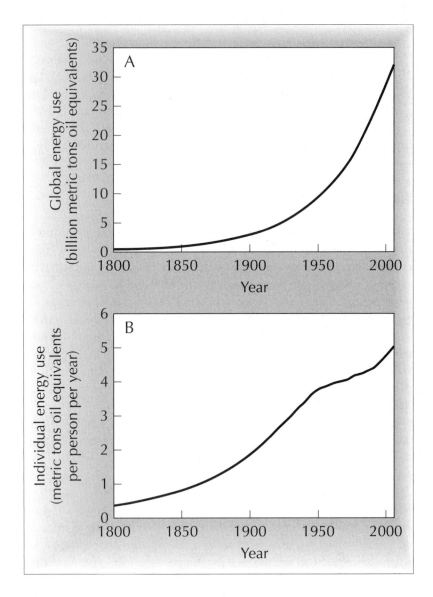

FIGURE 2.3 Global energy use from 1800 to 2005, expressed in terms of oil equivalents: (A) total energy use is increasing, as is (B) the average amount of energy use per person. Total energy use includes the major forms of energy used converted into the mass of oil required to produce that energy.

horsepower. A healthy human can sustain about 0.1 horsepower. In ancient times, only kings and emperors could afford to have 200 horses (or 2000 to 3000 people) to carry them and their possessions from place to place. Now, most adults in developed countries and many in other countries can control that much power if they buy or rent a large vehicle. As more of Earth's population attains a lifestyle similar to that enjoyed in developed countries, average per capita energy use will increase. Increased per capita energy use means greater global energy use, unless substantial improvements in efficiency of energy use occur. Given that both standard of living and total number of people are increasing, energy use will probably increase even faster. Thus global energy use rates will continue to increase, even if energy use rates stabilize in developed countries. The past global trend has been for increased energy use, and it will continue into the future. Huge strides in energy use efficiency, including more sustainable power sources and lower use rates, will be required to offset global environmental impacts.

Using Up the Freshwater

Water is a resource for which there is no substitute; all life depends on it, and human civilizations require abundant and reliable sources of freshwater. Modern economies require ample supplies of water to sustain them. Close to 1 billion people currently lack access to safe drinking water; polluted water contributes to deaths of about 15 million children under age five each year. The United Nations predicts that in 45 years, some 2 billion to 7 billion people will face water shortages.

Few people realize how much water is required to sustain lifestyles in developed countries. Each hamburger requires 800 gallons (3028 L) of water to produce. A full meal with deserts and drinks (including 530 gallons [2017 L] of water to produce a glass of brandy) requires 1477 gallons (5591 L). It takes about 25 bathtubs of water to produce the cotton in one T-shirt.[13] An acre of corn requires 500,000 gallons (1.8 million L) of water. Global water use rates have grown exponentially over the past 200 years (figure 2.4). Mirroring trends in total global domestic product and energy use, water use has increased by 500 percent in the past 100 years. The picture is somewhat brighter than that of energy use: per capita water use rates peaked around 1950 and have decreased about 15 percent since then. Unfortunately, in recent years, the rate per person

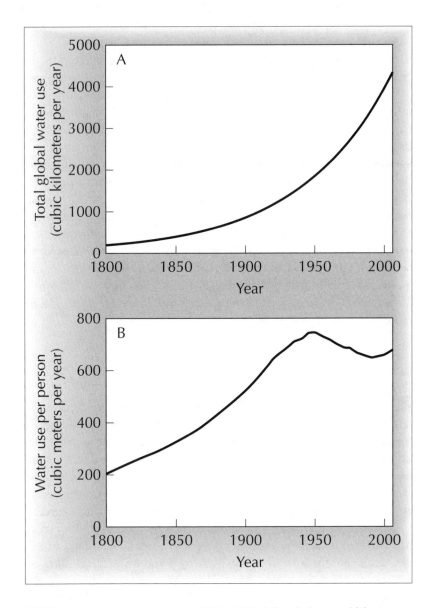

FIGURE 2.4 Human water use from 1800 to 2005: (A) global use and (B) use rates per person. The total amount of freshwater available to humans is about 2100 cubic miles (9000 km³) per year. Humanity is using about one-half that amount now, with an annual rate of use per person of 179,000 gallons (678 m³).

has started to increase again. Why might the expansion of water use be a problem if it continues?

The hydrologic cycle (movement of water across Earth—evaporation from oceans, to precipitation, through rivers, and back to oceans) continuously renews water for human use. Only a set amount of water (200 cubic miles per year, or about 9000 km^3 per year) falls as precipitation in areas, or at times, in which it is useful to humans.[14] A substantial amount of precipitation that falls on land falls in areas far from human population centers, such as in the Arctic and Amazon basins. When it rains, it pours; floods move large amounts of water past areas of human inhabitation so quickly that it cannot be used. Evaporation continuously returns water used for irrigation or stored in reservoirs to the atmosphere. Pollution can make water unusable. This water cannot be captured for human uses.

Humans rely on water directly for irrigation (irrigated crops provide about 40 percent of the world's human food) and for domestic and industrial purposes. About 70 percent of global water use is for irrigation of crops. We rely on water indirectly for removing and treating wastes, to provide habitat for aquatic organisms we consume, for powering industry, and to generate electricity. In addition, many plants and animals rely on freshwater for part or all of their lives. Freshwater is a shared resource and has the potential to become limited on a global scale in our finite world.

Many benefits of water exist in addition to direct uses, including fish and wildlife habitat. Water also facilitates vital functions of waste purification and removal for human society. Aquatic habitats provide about $28 trillion per year in economic benefits across the globe: $21 trillion from marine habitats, $5 trillion from wetlands, and $2 trillion from lakes and rivers. These beneficial services include climate regulation, nutrient cycling, flood regulation, food production, recreational uses, cultural values, and waste treatment.[15] To put the value of water in an economic context, the value of aquatic resources per year is roughly equivalent to global economic output listed in the preceding section on economic growth. While these exact numbers are questioned by some economists, any economic analysis of global human impacts on our aquatic resources should not discount the value of maintaining abundant clean water.[16]

The city of New York has invested $4.5 billion in restoring watersheds in the Catskill Mountains, and this investment could forestall building massive purification plants.[17] Development in the Catskills was degrading water quality, and a $6 billion purification plant was going to be needed to provide drinkable water to the city. Repairing old sewage plants and protecting the natural ability of watersheds to provide pure water from intact natural systems was the most cost-effective option. Given the value of water to New York City, consider how much it is worth to supply water to all large cities on Earth.

China is water limited in the north and wetter in the south. The Chinese are now starting a $62.5 billion project that will include canals and pumping stations to transfer water to the north. This project is required to meet the projected water needs of almost 0.5 billion people. The exact environmental impact of such massive interbasin water transfers is not certain, but there are worries that polluted water from the south will contaminate the north and that dams and canals built to feed the water-transfer system will damage Tibet's fragile watersheds.[18]

One temporary solution to water scarcity is to pump up groundwater. About 20 times more fresh groundwater exists within about 3 miles (4.8 km) of Earth's surface than is on the surface in lakes and rivers at any one time. The shallow portion of this water that lies under agricultural areas is already heavily exploited for irrigation around the world.[19] In such regions, supply is often used at greater rates than groundwater is replenished from rainfall and percolation from above. The deeper underground water is located, the higher the energy costs to pump it to the surface. Costs to pump water to the surface must be less than benefits water can provide, or it is not an exploitable resource. For example, if costs of pumping groundwater make it impossible to make a profit from farming cropland, then groundwater is not a reasonable agricultural resource. We are essentially mining groundwater because we pump it at unsustainable rates, particularly in arid areas where water is most limiting.

A prime example of mining groundwater is the Ogallala (High Plains) Aquifer, which underlies parts of Nebraska, Colorado, Kansas, Oklahoma, Texas, South Dakota, Wyoming, and New Mexico. About one-quarter of the cropland in the United States is watered from the Ogallala. Water use of 1.7 billion gallons (6.4 billion L) per day across the aquifer has lowered the level of groundwater 40 to 60 feet (more than 10 m) in many regions. The rivers and streams that once flowed across the Great Plains are mostly dry, and species of fishes are disappearing. Once the groundwater is gone, natural rainfall rates will not replenish it for centuries. Similar withdrawals are common worldwide.

Oceans contain tremendous amounts of water. Why don't humans simply use this source to provide for their needs? Making saltwater useful for most human needs requires desalinization, which necessitates substantial amounts of energy. Water is naturally desalinated when the sun heats the ocean and causes evaporation. It takes 540 calories to evaporate about one-fifth of a teaspoon (1 g) of water. Energy required for evaporation is a fixed amount determined by physical laws; no amount of scientific progress can reduce the amount of energy required to evaporate water. Solar energy required to evaporate the 200 cubic miles (9000 km^3) of precipitation that enters the atmosphere from the ocean is about 58 times the total human global energy use. Roughly 23 times the current annual human global energy use would be required to evaporate the amount of water we now use globally.

An added energy cost of desalinated water is transportation. Water is very heavy per unit volume and is extremely costly to transport. Just try to carry two full buckets of water up a flight of stairs. Water in the hydrologic cycle is lifted into the atmosphere and transported by wind moving clouds onto continents. The tremendous amount of energy in water that is evaporated into the atmosphere and carried to land by weather patterns is partially exploited by humans as hydropower-driven electric generators. The energy it takes to lift water in natural precipitation comes from solar heating of the atmosphere. The average elevation of Earth's surface is more than 0.6 mile (1 km). To lift the 200 cubic miles (9000 km^3) that are available for human use to the average elevation would require approximately 20 percent of total energy used by humans on Earth each year. This is the same amount of energy it would require to pump an equivalent amount of groundwater from 0.6 mile (1 km) belowground up to Earth's surface.

Energy to lift water involved in the hydrologic cycle does not include energy it would take to move water from oceans across continents—horizontal movement, as opposed to vertical movement. The energy to move water laterally is more difficult to calculate, so I leave it to the reader to imagine the trucks, trains, or giant pipelines with pumps that would be required to move desalinated water from the ocean to where it is needed. An everyday example of the cost of transporting water is the high cost of bottled water relative to tap water. The water itself in a bottle of drinking water costs only pennies; the cost (above profit) is due to the considerable energy costs of transporting bottles of water.

Given all the energy required to purify and move water in substantial amounts, desalinization is only practical for areas near marine coasts. Uphill transfer of water from one river basin to another is very expensive. The problem is not how much water is on Earth but when and where it falls as precipitation, and, once it falls, how much it has been polluted by humans.

Water shortages can lead to political turmoil because water is limiting and valuable. More than one-third of all people on Earth live in areas where water is in short supply, and 1.7 billion people reside in regions where chronic water shortages hinder crop production and economic development.[20] Some 1 billion people (about one out of six) currently lack adequate clean drinking water.[21]

As world population increases and water is used more heavily, pressures on an extraordinarily valuable resource with a finite rate of supply will intensify, as will political problems related to water supply. The fact that humans already use one-half of the readily available water is a clear indication that there is a practical limit to increasing rates of resource use. For example, worldwide water use more than doubled in the past 50 years. If water use doubles again in the next 40 to 50 years, *none* will remain for anything other than direct uses such as drinking.

Either golf courses and swimming pools or species living in the few rivers that flow through dry areas (such as Phoenix and Las Vegas) will become extinct. The propensity of humans for luxury consumption is high, so I would not bet against golf courses. Worse, in lesser-developed countries, rates of malnutrition and starvation could increase if adequate water is not available for crops. Water availability is a serious issue, and current trends in use obviously cannot continue to be supported.

Water availability and quality issues also can be used to illustrate complex interactions among different facets of human impacts on the environment. Scientists predict that summer runoff will decrease as temperatures increase and that water demand will increase in response to increased carbon dioxide and the greenhouse effect (chapter 3).[22] These two factors will conspire not only to degrade water quality but also to decrease water quantity,[23] particularly in parts of Europe, the Middle East, and the western United States.[24] Clearly, a complex web of interactions at the global scale requires deep scientific understanding of the hydrologic cycle and how humans alter it.

Will we reach a point where the only flow in rivers of dry areas with moderate to high human population density comes from sewage-treatment plants and pipes draining cropland? We are already there in the Great Plains. The Arkansas River historically flowed from the Rocky Mountains across Kansas year round; now the river is dry where it passes over the Ogallala Aquifer, and the headwaters of the Arkansas River in Kansas often is the Great Bend sewage-treatment plant. Sewage and agricultural runoff provide all stream flow in many arid regions today. In arid regions, water is used repeatedly as it passes downriver. If we use water intensively before returning it to streams, huge amounts of money will be required to make it fit for drinking, supporting aquatic life, or irrigating crops. If all surface water is polluted or stream channels are dry, it will take a tremendous toll on freshwater species and the ability of aquatic ecosystems to function in ways that are beneficial to humans.

There is some reason for optimism; a recent analysis indicates that we could manage water sustainably, to support water supplies and ecological integrity, with existing technology.[25] This analysis also notes that if human water demand increases, such sustainable use will become less likely.

Transforming Earth into a Farm

The percentage of Earth's surface covered by cropland has expanded greatly in the past 300 years. The rate of expansion of cropland area has slightly decreased in the past few years, and the trend over the past century is almost linear (not exponential, like the global trends shown previously). When area of cropland is divided by number of people, total area of cropland per person has decreased by about one-third in the past 50 years (figure 2.5).

Cropland is providing more food per acre since the 1950s because of "green revolution" technology, which leads to greater crop yields due to irrigation, mechanization, fertilization, pesticide use, and improved genetic strains of crops. For example, expanded use of irrigation has increased crop yields. When water becomes limiting (as suggested by extrapolating the data in figure 2.4), further improvements in yield by increasing the amount of irrigated land will not be possible. Productivity increases are also related to more mechanized techniques (which require greater energy input), improved crops, and extensive application of pesticides and fertilizers. All improvements in crop yield require large inputs of energy and have corresponding environmental effects

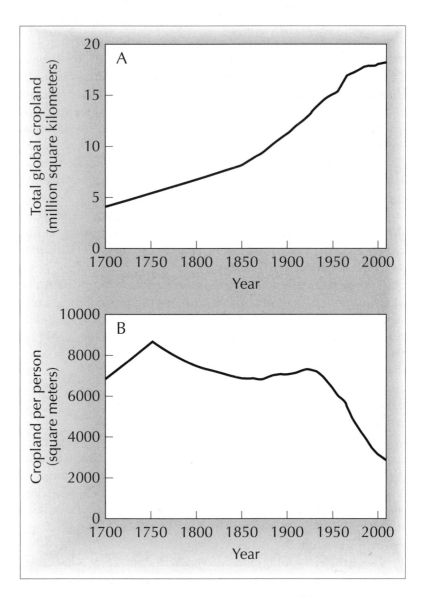

FIGURE 2.5 Average cropland from 1700 to 2005: (*A*) global area and (*B*) area per person, showing that available cropland may not accommodate future increases in human population unless crop yields increase substantially.

(some of which are described in chapter 3). Currently, there are more malnourished people than ever before in history. Adequately feeding all people now on Earth would require reforming the institutional and economic impediments to relieving hunger (although we currently have enough agricultural capacity to nourish all people, we allow many to be malnourished or starve). Even if the institutional and economic impediments to feeding all people were removed, more cropland or greater yields per unit area of cropland would be needed as human numbers and demands increase.

Our ability to increase the amount of food produced to match human population requirements is not certain, because both population and standard of living (increased food consumption) are expanding globally. Increases in standard of living substantially raise crop consumption because animal protein becomes more important in the diet of people with a higher standard of living. Creation of every 1 pound (0.45 kg) of animal protein requires 8 or more pounds (3.6 kg) of grain protein. Eating animal protein uses more cropland than obtaining protein directly from grains. Thus people are putting greater demands on cropland production because both population and rate of per capita resource use (in this case, crop resources) are increasing. As people in developing countries obtain the resources to afford a more protein-rich diet, absolute demand for cropland will continue to expand.

Much agricultural production is directed toward luxury products and requires use of native habitats. For example, increases in the price of coffee stimulate deforestation in Sumatra, where the last remaining rainforests are under siege. The higher the price of coffee, the more lowland rainforest is taken over by farmers in "protected" national parks in Lampung.[26] While many would argue that coffee is essential, particularly in the morning, few realize the ecological damage inflicted by coffee farmers who are simply trying to eke out a living.

Tremendous increases in productivity of cropland made possible by technology in the past 50 years may not be duplicated. Most high-quality areas for growing crops are already under cultivation. Increasing food production will take a large environmental toll. It will be a challenge to feed humanity while maintaining the ability of Earth to support humanity in other ways. Converting natural lands to croplands endangers species, water supply, and water quality.[27] As we demand more production from marginal lands, wind erosion increases, thereby leading to air pollution so that devastation similar to that of the Great Dust Bowl in the United States will occur. (I discuss some specific environmental problems associated with agriculture in chapter 3.) Feeding the world's population is one of the great challenges for now and the future, and the larger Earth's population, the less likely it is we can meet this challenge.

Can Amplification of Resource Use and Population Growth Continue Indefinitely?

All global trends related to human population and resource use are increasing explosively. Human population, world gross domestic product, energy use, and water use are all increasing exponentially or faster. Rates of use in each decade, for the most part, have exceeded those in the decade before. What price will we pay for burgeoning of rates of resource use, and can such expansion continue? No prediction of the future is guaranteed to be accurate, but I find it difficult to look at increasing population and resource use rates and corresponding environmental effects without being concerned.

Long-term trends in population growth and resource use confirm the idea that human behavior has been predictable for the past several hundred years. Had we analyzed trends between 1900 and 1950—trends in human population numbers, economic activity, energy use, area of cropland, and water use—and extrapolated the same rate of expansion to the period 1950 to 2000, our predictions would have been fairly accurate, and in some cases we would have underestimated current rates of resource use and population growth. From 1900 to 1950, human population increased 1.9 times; gross domestic product, 2.7 times; world energy use, 4.9 times; water use, 2.3 times; and area of cropland, 1.4 times. From 1950 to 2000, human population increased 2.4 times; gross domestic product, 4.3 times; world energy use, 3.1 times; water use, 3.8 times; and area of cropland, 1.3 times. All these expansions have been occurring for several generations and show little sign of halting in the near future. Can Earth support the same amount of an increase over the next 50 years? Can we support 10 billion people at a standard of living found in developed countries?

Of the resources previously described, energy use will not be directly limiting for lack of fuels, because large amounts of other energy resources—nuclear, wind, hydroelectric, and solar energy—have not been fully exploited. Greater energy use will probably lead to greater exploitation of sources of fossil fuels such as coal, oil-sand, oil-shale, and methane deposits in the next 50 years as reserves of crude oil deposits are depleted. These alternative fossil fuel sources will be more expensive to use and, in many cases, more damaging to the environment. Using them without increasing the rates of pollution will cost even more. Each alternative energy source has its attendant environmental impact: hydroelectric power interferes with natural river flows, nuclear waste for political and technological reasons is very difficult to dispose of safely, nuclear plants are not always safe for the surrounding population (for example, the explosions at Chernobyl and Three Mile Island), and production of solar panels uses toxic chemicals. Managing impacts on the environment from human energy use will require careful consideration of benefits and drawbacks of each alternative source of energy.

Amounts of cropland could limit the standard of living of many people in the future. Much of the best cropland on Earth is already being used to produce food, and soil erosion and urbanization are claiming some of the best cropland. Many areas that could be irrigated are already under cultivation, and groundwater for irrigation is being used up. Yet the expansion of cropland and yields is necessary to satisfy the demands of Earth's future population.

Limits to total water availability are clear-cut. Water shortages can have many negative consequences beyond the obvious limitations on domestic use. Water is used, among other things, for irrigation, industry, power generation, and transportation. Pressures on water supply as a resource are great, and prospects for increasing supplies are not good. At a minimum, it will take tremendous amounts of energy to make more water available to humans because of the energy it takes to move water from areas of high availability to those with low availability and the energy required to desalinate, purify, and transport ocean water.

One could argue that data behind the trends discussed so far are too soft and that large errors are inherent in estimating anything on a global scale. No one familiar with the issues disagrees with the trends until now; disagreement is over what the data mean for the future. Those who dispute the data's implications may point to uncertainties in order to claim that environmental protection "is not worth it." Even if estimated trends are off by 20 percent (one-fifth), their direction is clear. When trends of resource use and population growth demonstrate tenfold or greater increases over the past 200 years, it does not matter if the absolute number at any point in time can be determined only plus or minus one-fifth of the value. In spite of inherent uncertainties in any large-scale accounting, trends are real. The issue is interpreting what trends mean to humans.

Some argue there is *nothing* bad about growth of economic activity on Earth. An improved standard of living for more people is highly desirable; poorer people should experience the same well-being as those fortunate enough to inhabit economically advantaged parts of the developed world. But there is no proof that sustainable inhabitation of Earth is possible if human population growth and ever-increasing appetites for resources continue.

Humanity's Footprint

Unlimited expansion of human population on Earth is a physical impossibility. Calculating how many people we can pack onto Earth's surface is an academic exercise; a more realistic approach is to calculate how much resource (for example, surface area of Earth, energy, water, food, habitat for waste purification, atmosphere to absorb pollutants) it takes to support each person, and how much of the natural world we want to preserve for purposes other than supporting humans (for example, national parks and nature preserves for their recreational and aesthetic values). The area required to support each human is referred to as an "ecological footprint." The average ecological footprint in the United States is about 25 acres (10 ha), and the worldwide average is 6.7 acres (2.7 ha).[28]

Recent detailed analysis demonstrates that human population size and affluence are the primary drivers of the human ecological footprint.[29] This analysis suggests that other factors—such as urbanization, economic structure, age distribution, life expectancy, and degree of education—have little influence on the ecological footprint. These researchers predict that the human ecological footprint will increase by about 33 percent over the next decade, and we need to increase our efficiency of resource use by 2 percent every year to counter the increase in footprint.

Intensity of impact can also be calculated. There is no place on the surface of Earth where no ecological footprint is present, but some areas have very intense human influence related to high population density or resource use rates. The areas with the largest local footprints occur in Southeast Asia, Europe, and the eastern United States.[30] In some places, uses far exceed capacity for support, and these areas must receive inputs of materials from other areas.[31] Calculating how many people we can support while still maintaining Earth will be necessary if we want to provide a decent quality of life for future generations.

More detailed calculations of human influences on the Earth that supports us extend the ecological footprint concept.[32] These calculations account for six components: growing crops; grazing animals; harvesting timber; fishing; accommodating housing, transportation, and industry; and burning fossil fuels. Calculations indicate that human population exceeded Earth's capacity to support humanity indefinitely in the 1980s. By 1999, we were at 120 percent of the globe's capacity for sustainable support. The idea of exceeding the sustainable footprint has been termed "overshoot."[33] We are overshooting the ability of Earth to support us sustainably.

3

Shock Wave from the Insidious Explosion

Momentum of Human Effects on Our Planet

Global increases in population and resource use are creating a shock wave that is ripping through the global environment supporting humanity. Just as a large explosion creates a shock wave that carries the explosion's power far from a detonation site, momentum of resource use rates is extensive and has spread to all parts of Earth, even those far from human habitation. Such broad global impact is unique in human history. Human changes to the environment are occurring at such a large scale and with so much momentum that reversing them will take at least decades, but probably centuries, even if we act now. The capacity of humans to overuse their environment has been recognized for decades; still, only 25 years ago, the greenhouse effect, ozone depletion, global species extinctions, worldwide deforestation, global transport of toxic compounds, and the finite nature of water supplies were not commonly mentioned in environmental texts.[1] Today we know how humans influence the Earth they have filled up. It makes sense that actions such as those that can change the temperature of the entire Earth will take a considerable amount of time to reverse, due to the global scales that impart tremendous momentum.

Evidence for Humans Damaging the Local Environment
That Supports Them

Humans have had strong local influence on their environment since ancient times. Over the past 50,000 years, humans have caused extinctions of most large animals in Europe, Asia, North America, and Australia.[2] For example, when humans reached Australia, they increased the rate of fires, shifted the landscape to a fire-adapted desert shrub, and simultaneously hunted large animals. These human pressures ultimately led to the extinction of 60 large animal species.[3] This removal of large herbivores almost certainly influenced plants and other animals in the ecosystem. Other human activities, such as deforestation, also strongly altered local environments. In some cases, alteration substantially reduced the ability of local ecosystems to support human civilizations, and some of these impacts were severe enough to cause the decline of entire societies. Historians and anthropologists hypothesize that the Mayan, Anasazi, Mesopotamian, and ancient Greek cultures declined because of unwitting, self-inflicted environmental destruction.[4]

One of the most dramatic cases of humans harming the ecosystems that support them is that of the Polynesians who colonized Easter Island some 1500 years ago. The early inhabitants quickly developed a social structure that was capable of creating large stone monuments. Anthropologists offer two explanations of what followed.[5] In the first scenario, population expansion, coupled with resource use required to support the population, eventually caused the collapse of the civilization. This collapse may have been too quick for individuals to realize it was happening in time to reverse it. The scenario blames humans for deforestation, making it impossible to build the seaworthy canoes that were necessary for trade with other islands that had resources they needed. Lack of good watercraft and no extensive nearby shallow reefs became problematic for a people relying in part on seafood taken from open waters. The modest number of nearby shallow marine habitats initially provided food, but the large fishes that inhabited shallow waters and nearby reefs were rapidly depleted by increased hunting pressure. The inability of the island to support the population is thought to have led to warfare, population decline, and cannibalism.[6] In this case, the population simply grew too large for the island to support. People's demand for resources overcame the environment's capacity to provide them, leading to societal collapse and imposition of natural constraints on population size. The second scenario puts the blame for deforestation in part on Polynesian rats introduced as food by early inhabitants. The rats ate nuts from palm trees that dominated the island, leaving the palm trees unable to reproduce. In addition, the use of wood by people led to almost complete deforestation of the land. In this scenario, the island's ability to support an increased population was destroyed before the population could grow very large.

Both scenarios are examples of people unwittingly diminishing the provision ability of their environment. Regardless of which scenario transpired (or a mixture of the two), Easter Islanders drove at least six species of birds to extinction and caused the extinction of 21 species of plants that they used for food or materials. All of the monument building and environmental destruction was accomplished with primitive tools; the inhabitants of Easter Island had no machines. Modern humans can take a much higher per person toll on the environment than did the inhabitants of Easter Island. The Easter Island example is particularly instructive because once the trees were removed and seaworthy canoes could not be made, the living area was finite. Earth also is finite with respect to the current massive human population and resource use rates. Easter Island is a miniature example of what we are facing globally today.

We have entered into a new era in human history: that of a species with the capacity to serve as global environmental engineers. Over the past 50 to 100 years, we have escalated environmental change on a global scale.[7] The industrial revolution accelerated global change that had started with the spread of agriculture.[8] This level of global change caused by organisms has occurred only several times in Earth's 4.5-billion-year history, most notably when cyanobacteria (numerous species of primitive blue-green algae) oxygenated the atmosphere almost 2 billion years ago.

The current level of global environmental impact caused by a single species is unprecedented in Earth's history. Unlike in ancient times, in our world there is nowhere to go to escape these impacts if they are of a great enough magnitude to destabilize or severely hinder the capacity of Earth to support our current civilization. Global environmental effects include (but are not limited to) increased atmospheric carbon dioxide and the greenhouse effect, ozone depletion that increases ultraviolet radiation, water contamination associated with agricultural chemicals, introduction of exotic species, and habitat destruction that causes species extinction. I discuss the first three effects in this chapter and the last two in the next.

Carbon Dioxide and the Greenhouse Effect:
Poking the Monster with a Stick

Two of the strongest indicators of human impacts on global environment are increase in carbon dioxide (figure 3.1) and decrease in ozone
in our atmosphere. Carbon combined with oxygen, nitrogen, and hydrogen make up all cells. When this carbon in cells is metabolized, or
when it is burned, the other elements are stripped away and the carbon
combines with two oxygen atoms to make carbon dioxide (CO_2). Carbon dioxide in our atmosphere has increased by one-quarter over the
past 200 years. The increase in CO_2 is exponential, and scientists have
directly tied it to human activities.

Carbon dioxide is a gas produced by combustion of carbon (contained in wood, coal, and oil); by volcanic activity; and by respiration,
the metabolism of all animals, microorganisms, and even plants. Carbon dioxide is removed from the atmosphere by photosynthesis of the
trees, herbs, and grasses on land and by microscopic plants in oceans.
Photosynthesis provides carbon that makes up the mass of almost all

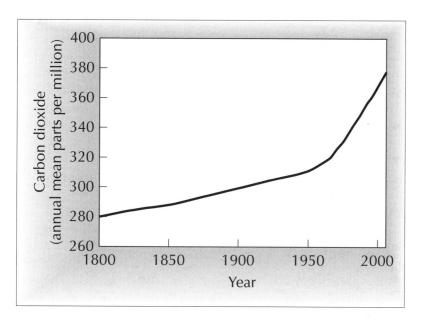

FIGURE 3.1 Concentration of carbon dioxide in Earth's atmosphere from 1800 to
2005, showing an increase of close to 30 percent.

organisms on Earth. Land plants photosynthesize, and some carbon that is turned into plant tissue is deposited into soil and stored. Carbon dioxide absorbed into oceans is used by microscopic photosynthetic plants to produce carbon-containing compounds that make up their cells. Some of this carbon sinks and eventually is deposited on the bottom of the ocean.

Natural processes involved in CO_2 production and storage have kept the balance of CO_2 relatively constant for much of human history. Fossil fuels represent a large amount of fossil carbon-rich plant material that was deposited and stored underground over millions of years. A portion of CO_2 taken up by plants is deposited in soil or sediments when they die and are buried. Deposited materials form oil, gas, and coal reserves underground. Over millennia, these fossil fuels have come to contain a tremendous amount of carbon. During Earth's geologic history, volcanic activity slowly has been releasing CO_2 into Earth's atmosphere from its crust. This volcanic release would increase atmospheric concentrations of CO_2, except that plants use it for photosynthesis and their remains are buried, resulting in a balanced CO_2 concentration.

Human uses of fossil fuels have disrupted this balance by drastically increasing the pace at which stored carbon is released into the atmosphere as CO_2. Starting with the industrial revolution in the nineteenth century, energy use has increased radically (chapter 2). The industrial revolution and attendant expansion of the global economy over the past 100 years has been based on burning substantial amounts of fossil fuels. Moreover, forests store large amounts of carbon, and deforestation disrupts the balance of atmospheric CO_2. Plowing land for crops encourages carbon in the soil to return to the atmosphere as CO_2. In the past several thousand years, humans have removed large areas of forests, and this deforestation may have already started altering our climate. Rates of deforestation have accelerated over the past few hundred years, thereby increasing the global effects of CO_2.

Increased use of fossil fuels and rates of forest destruction have ultimately led to an exponentially increasing concentration of atmospheric CO_2 gas. Between the end of the nineteenth and the beginning of the twentieth century, most of the increase in atmospheric CO_2 was due to deforestation and agricultural practices; after that, the effect of burning fossil fuels eclipsed the effect of deforestation. Now, the use of fossil fuels causes more than three times the CO_2 emissions accounted for by continuing global deforestation.[9]

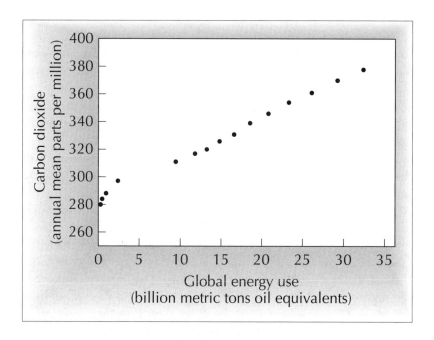

FIGURE 3.2 Relationship between global energy use and atmospheric CO_2. Although correlation is not causation, evidence is compelling that humans' use of coal, oil, gas, and wood as energy sources has increased CO_2 inputs to the atmosphere. This, coupled with the destruction of forests that serve as CO_2 storage sinks, has led to the increase of CO_2 in the atmosphere.

Since 1800, the concentration of CO_2 has increased at a rate 10 to 100 times greater than any time in the past 500,000 years.[10] This increase is directly correlated with the amount of energy we use on Earth (figure 3.2). The correlation is strong because oil (including gasoline), coal, and natural gas, deposited over millions of years, is being burned at ever-increasing rates and releasing CO_2. Carbon dioxide is released from burning fuels at a rate of 6 million trillion tons of carbon per year, and the remaining emission of CO_2 is currently due to the clearing and burning of forests.[11]

The rise in CO_2 will be evident for all of our lifetimes. Even if we immediately halted the release of CO_2 (something that clearly will not happen), the effects on Earth's atmospheric chemistry would remain for at least 100 years.[12] The time required for the atmosphere to recover from CO_2 inputs is yet another example of momentum of the "global environmental catapult"; decades or centuries are required for global processes to reach equilibrium (that is, return to a stable state).

Carbon-capturing processes of marine and land plants are slow, so it will take at least decades for excess built-up atmospheric CO_2 to be removed and buried in soil or ocean sediments.

Even though CO_2 makes up less than 1 percent of gas in the atmosphere, it has a major role in regulating global temperature. Carbon dioxide absorbs heat energy and traps it in Earth's atmosphere. The sun heats Earth, and Earth radiates energy back into space. When some of that energy is trapped before radiating back to space, the atmosphere gets warmer. This increase in temperature is called the "greenhouse effect" because Earth warms up just like a greenhouse that traps solar heat. If CO_2 concentrations increase, more energy is trapped, and atmospheric temperatures increase. The greenhouse effect was first noted in the 1890s by Scante Arrhenius, a Swedish Nobel Prize–winning chemist. Being from a cold country, he thought warming was good because it would make Swedish winters less severe.[13] By 1903, Alfred Russel Wallace (co-discoverer of the theory of evolution by natural selection) had documented the delicate balance of Earth's temperature, CO_2, and global temperature.[14]

Evidence is overwhelming that the increase in CO_2 and some other gases (for example, methane and nitrous oxide) associated with human activities (such as burning fossil fuels and deforestation) is causing a gradual warming of Earth. Within the mainstream scientific community, the chief controversy revolves around how much warming will occur and what environmental effects warming will have. The American Geophysical Union is the world's largest society of Earth scientists, with 31,000 members, at least 4000 of whom are atmospheric scientists. In 2003, the American Geophysical Union adopted a resolution stating that increases in atmospheric concentrations of CO_2 and other greenhouse gases will cause the global surface climate to be warmer.

The Intergovernmental Panel on Climate Change (IPCC) is composed of experts from around the world on climate and factors influencing the climate and its relationship to the environment. In its "Third Assessment Report," issued in 2001, the IPCC expressed a 90 percent certainty that global temperatures will continue to rise, and the National Academies of Science of 14 countries, including the United States, have recognized the scientific validity of the panel's findings.[15] The IPCC and Al Gore won the Nobel Peace Prize in 2007 for their independent work on global warming. The IPCC's 2007 report includes a 90 percent certainty that humans are causing global warming:

The understanding of anthropogenic warming and cooling influences on climate has improved since the Third Assessment Report (TAR), leading to *very high confidence* that the globally averaged net effect of human activities since 1750 has been one of warming. Warming of the climate system is unequivocal, as is now evident from observations of increases in global average air and ocean temperatures, widespread melting of snow and ice, and rising global average sea level.[16]

The few scientists who disagree with the vast majority about global warming garner public attention because controversy generates news stories. Promulgation of this minority scientific view is encouraged by those who do not want to curtail current usage rates of fossil fuels. People, companies, and governments with a financial stake in maintaining or increasing rates of CO_2 releases latch onto the scientific views of a tiny minority to justify not curtailing continued increases in rates of release of greenhouse gases.

The IPCC report of 2001 estimated a global average temperature increase of 1.0 to 4.5°F (0.6 to 2.5°C) by 2050 and 2.5 to 10.4°F (1.4 to 5.8°C) by 2100.[17] The 2007 report estimates almost a 1.8°F (1°C) temperature increase since 1960 on land and confirms the projections of the 2001 report. It warns that continued increases in release rates of greenhouse gases will exacerbate global warming. Data are accumulating from numerous sources confirming that global temperature increases are real (that is, the rise in global temperature in the past century is highly statistically significant) and are influencing both the biological characteristics of Earth and the physical characteristics (for example, climate and atmospheric and oceanic temperatures).

Biologically, a 95 percent certainty exists that heat trapped by greenhouse gases has warmed the oceans.[18] Temperature of soil and rock on all continents (Earth's surface crust) except Antarctica increased slowly between 1500 and 1900 but has increased sharply since 1900.[19] Since ground temperature is driven by air temperature, it provides strong evidence for a recent increase in atmospheric temperature in the past 100 years, compared with the previous 300 years. Physical signs that Earth's temperature is increasing include a 6- to 8-inch (10- to 20-cm) rise in sea level in the past century, warmer ocean temperatures,[20] decreases in Arctic and Antarctic sea ice (see figure 1.1), and decreases in glaciers around the world.[21] When Glacier National Park was first visited by naturalists around 1900, there were 150 glaciers. Now there are only 35, and none will remain in approximately 25 years. Loss of glaciers and snowfields endangers water supplies in some areas such as India and South America.[22]

These global effects have considerable momentum. Global climate models show that if concentrations of greenhouse gases were stabilized back down to 2000 levels (which is not going to happen soon), there would be at least an additional 1°F (0.6°C) increase in atmospheric temperature. Since ocean water expands when heated, atmospheric temperature increase would lead to a sea level rise of up to 4 inches (8 to 10 cm) over the next 100 years. The most conservative estimates suggest another 2-inch (5-cm) increase due to melting of glaciers and ice caps on land over the next century.[23] Warming is also increasing the rate at which glaciers flow, and, as a result, sea level could rise more than current forecasted levels, which underestimate the input from glaciers in Greenland and Antarctica.[24] With increasing global greenhouse gases, more dire predictions are that sea level could increase as much as about 1 foot (30 cm).[25] The poorest, low-lying countries will be most vulnerable to rises in sea level because they are unable to afford extensive sea walls and levees.

Other consequences of warming are more difficult to predict. Although weather prediction loses accuracy the further into the future predictions are made, it is reasonable to assume that higher temperatures will cause further rises in sea level, changes in biology of plants and animals, extinction of some desirable species, greater spread of human diseases, and expansion of pests into habitats that were formerly too cold for them. About one-quarter of all species on the planet may become extinct by 2050 from the effects of global climate change.[26]

Species all over the world are exhibiting changes in their behaviors and distributions as a result of global warming.[27] For example, British plants flowered 15 days earlier in the 1990s than in previous decades.[28] Change in flowering time, in turn, changes relationships between plants and their pollinators. Since we rely on pollinators for a significant portion of our food crops, direct implications of global warming for human food production related to indirect effects on many species are possible. Pollinated food crops include most fruits, cucumbers, squash, tomatoes, alfalfa, nuts, coffee, dates, figs, cacao, and others.

Copepods, small crustaceans in the same group as shrimp, are a vital part of the food base for marine ecosystems. Copepods consume microscopic plants and are eaten by small fishes. They are an integral food source in important marine fisheries. They excrete fecal pellets that move nutrients to the bottom of the ocean more quickly than nutrients would sink otherwise. Nutrients would otherwise stay in the surface waters and stimulate the growth of microscopic plants that ultimately provide food for fish. Increased marine temperatures have caused a northward shift of distributions of copepods by 10 degrees of latitude (roughly 600 miles, or 965 km). These changes in copepod distribution may stress already overfished areas by changing food availability to fishes and ultimately lead to alterations in basic ecosystem function.[29]

Total amount of precipitation on Earth will increase because more evaporation occurs when the atmosphere and oceans are warmer. Some areas, however, will experience less moisture as weather patterns shift; 12 climate models agree that there will be 10 to 30 percent decreases in river flow in southern Africa, southern Europe, the Middle East, and western North America.[30] Also, with temperature increases, evaporation is greater and soil dries more quickly, and water demand to keep plants alive in some areas will increase.

Storms and droughts will intensify because warming creates a more energetic atmosphere. Higher sea levels and more intense hurricanes and cyclones will probably cause increasing amounts of damage to human property and life. Tropical storm intensity is greater with warmer ocean temperatures. In 2005, there were more hurricanes than usual in the Gulf of Mexico, as well as the most powerful storms on record. The number of high-energy cyclones in the tropical Pacific and the proportion of dangerous category 4 and 5 cyclones have increased substantially over the past 35 years (figure 3.3). Damage from cyclones and hurricanes is exacerbated by a substantial portion of human population living in close proximity to coastal areas.

Consider the predicted temperature increase of 4.5°F (2.5°C) over the next several decades. In an area with an average August high temperature of 95°F (35°C), average temperature would increase to 100°F. Extended periods of temperatures exceeding 100°F (38°C) result in numerous deaths in major metropolitan areas. Global warming is predicted to intensify the frequency, intensity, and length of heat waves.[31] In 1980, a heat wave in the United States is estimated to have resulted in 1200 to 10,000 deaths; in 2003, an estimated 14,000 people in France and 35,000 people in Europe overall died from a single heat wave. While these heat waves may or may not be directly attributed to global warming, they illustrate how devastating heat waves can be. It has been suggested that deaths from heat waves are a "harvesting effect," culling

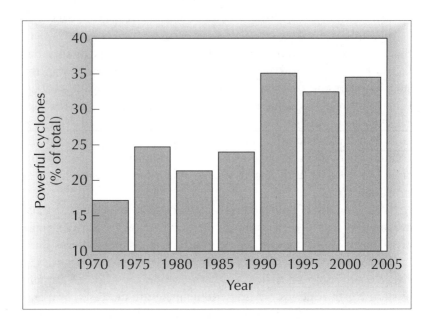

FIGURE 3.3 Average percentage of damaging cyclones and hurricanes world-wide, from 1970 to 2004. Although the total number of storms has not increased, the proportion of dangerously strong storms (category 4 and 5) has increased significantly. Storm strength increases with warmer ocean water temperatures (P. J. Webster, G.J. Holland, and H.-R. Chang, "Changes in Tropical Cyclone Number, Duration, and Intensity in a Warming Environment," *Science* 309 [2005]: 1844–1846).

from the population old and weak humans who have limited economic utility (yes, this really has been published).[32] But most people would agree that people dying from increasing temperatures is bad. Violent crimes in developed countries also escalate with warmer temperatures and higher heat indexes. The World Health Organization estimates that global warming already causes 150,000 additional deaths per year. Projections of the future effects of global warming on human health include increased mortality rates from diseases.[33]

Sea level rise will have profound effects on people living in coastal areas.[34] One effect will be billions of dollars in expenditures to protect low-lying areas in coastal countries; $200 to $400 billion in expenses is expected in the United States alone. Low-lying oceanic islands will be completely inundated. In Egypt, with about a 3-foot (1-m) rise in sea level, 12 to 15 percent of agricultural land would be at risk and up to 6 million people would be displaced without levees to hold back the sea. In Bangladesh, deaths already occur regularly from cyclones and related flooding from the ocean. In the 1970s, hundreds of thousands of people died from a single cyclone in the Bay of Bengal. Extreme weather events such as this will become more prevalent with global warming, and their effects will be intensified by a higher sea level. People in developing countries will suffer disproportionately because they do not have the infrastructure or resources to protect themselves from sea level rise and other aspects of climate change.[35]

The effects of warmer temperatures on our ecosystems will be considerable. Warmer climate has already caused more large forest fires in the western United States.[36] Weeds and disease organisms that are intolerant to frost will spread farther from the equator into temperate zones. This expansion will include infectious diseases such as malaria (carried by cold-intolerant mosquitoes). Malaria is already one of the worst human diseases in the world, and related diseases that infect humans and other species may spread with higher temperatures.[37]

Species lacking dispersal ability and high-temperature tolerance will become extinct. We have penned nature into parks and wildlife preserves; natural habitats are becoming ever more fragmented. In the past, plants and animals redistributed in response to global temperature changes such as those associated with ice ages. Past temperature changes occurred relatively slowly, so reproduction and dispersal allowed species to adjust their ranges. Humans disrupt natural dispersal pathways with urbanization, deforestation, road building, and other activities. Now many animals and plants cannot migrate or spread into more hospitable areas because there are few or no corridors of habitat left.

As temperatures rise, episodes of dissolved oxygen disappearing from lakes and near coastal ocean regions will become more severe, leading to declines in water quality and more frequent fish kills. Many types of marine life will die with only slightly higher temperatures. For example, corals have a relatively narrow range of temperature tolerance, and only a few degrees of warming can wipe out a coral reef. Declines of corals are consistent across the entire Pacific and Caribbean.[38] One-quarter of Earth's coral reefs may have already been destroyed or severely damaged by global warming[39] and human activities such as removal of large fishes.[40] Although these impacts are not likely to cause all corals to become extinct, widespread changes in coral communities, with sensitive species disappearing, are likely.[41]

A major concern with global warming is the potential for feedbacks that cause "runaway" warming. Such feedbacks occur when unprecedented global warming accelerates the release of greenhouse gases, and these releases cause additional runaway warming, which releases even more gases. The runaway feedbacks are particularly difficult to predict. The first feedback results from decreases in corals, which could contribute to runaway greenhouse gas production because coral skeletons are a major storage area for CO_2. Corals build reefs from calcium carbonate formed by chemical reactions of CO_2 dissolved in seawater. Increases in atmospheric CO_2 lead to increased acidity of seawater, which inhibits the formation of coral shells and related organisms.[42] Death of corals may lead to even greater CO_2 buildup because organisms that replace them do not deposit calcium carbonate originally taken from the atmosphere.

A second feedback could result from permafrost melting in the Arctic. Over one-third of carbon stored in soils on Earth is in frozen peaty permafrost areas of northern Europe, Asia, and North America.[43] As permafrost melts, soil organisms can process plant material stored in peat.[44] Like human respiration, soil organism respiration generates CO_2. Unlike human respiration rates, however, rates of respiration of soil organisms depend on external temperature: respiration rates are low when soils are frozen, then rise steeply with each degree of temperature increase. Stimulation of this processing of soil materials by increases in atmospheric temperature ultimately leads to greater rates of CO_2 release. Release of more greenhouse gases could cause even further warming, a feedback that would release CO_2 at even greater rates.

A third feedback from climate change could occur in the Amazonian rainforest. Deforestation makes forests more vulnerable to fire. Removal of the forest also leads to decreased rainfall in the Amazon because a lot of local precipitation comes from moisture released by tropical forests. Increased temperatures, coupled with fragmentation and decreased precipitation, will significantly increase fire rates. These factors could conspire to cause loss of a substantial portion of Amazon basin rainforest.[45] This large tract of forest is important for sequestering CO_2 because the trees take in CO_2 as they photosynthesize and grow. The loss of some of this forest could exacerbate increases in atmospheric CO_2, which could stimulate further forest losses.

A fourth feedback is the potential for catastrophic release of methane from sea sediments. Methane (natural gas) is stored in "clathrates," a combination of ice and methane. Large deposits are found in shallow areas of the Arctic Ocean. If the ocean warms, this gas could be released suddenly. Methane is a greenhouse gas, and a rapid release of a large volume of methane would lead to a corresponding increase in atmospheric temperature. Exact factors that will trigger these catastrophic feedbacks are not known, but the potential for devastation suggests it would be prudent to proceed with caution.[46]

Over centuries, other catastrophes related to large-scale climate change could occur. Increased river runoff into the Arctic Ocean could alter circulation in the Atlantic Ocean so that warm marine currents no longer heat northern Europe, leading to drastic local cooling. It would take almost a century for ocean currents to recover, and, in the meantime, agriculture in the region would be severely limited. Increased runoff into the Arctic Ocean has already started, likely as a result of global warming and climate change.[47] Researchers have a high degree of certainty that, over the next century, rises in ocean levels, harm to natural environments, and hotter temperatures will occur, but the rates and ultimate extent of these changes are less certain.[48]

Grimmer scenarios related to greenhouse effects lead me to worry about gambling with our atmosphere. Scientists who think Earth has the capacity to self-regulate its climate are concerned about perturbation of the atmosphere by humans, because they think processes that have regulated climate for millennia can be overridden by human emissions of greenhouse gases.[49] While it could be paralyzing to worry excessively about these negative environmental impacts, the potential for the feedbacks noted here should be addressed by policy makers worldwide.[50]

To extend an analogy attributed to respected climate scientist Wallace Broecker, the human experiment with our atmosphere is like poking an unpredictable monster with a stick.[51] This monster has been sleeping on a pile of gold for the past 1000 years. Legends are inconclusive as to whether the monster was violent or benevolent the last time it was fully awakened. Each time we poke the monster, it stirs a bit and we can take a little more gold. The gold has allowed us to build a large city around the monster's lair. Scientists note the monster's sharp teeth, human bones in the lair with teeth marks on them, and a lack of technology to fight the monster if it does wake up. Yet there are those who argue for poking the monster to take even more gold. Despite evidence and risks, those who stand to gain the most express the greatest certainty that there is nothing to worry about because poking the monster has not yet been proved to cause ill effects. They just keep poking the monster harder. Now we have reached the point where the monster has awakened for short periods and eaten one or two people. Still, some argue that it will be too costly to stop poking the monster.

Are short-term gains from our huge rates of energy use worth the risk they entail? The most prudent course of action would be to reduce greenhouse gas releases until greater predictive scientific certainty has been reached about the magnitude of temperature increase and its effects.[52] We already have all the technology to control emissions of CO_2 and meet Earth's energy needs over the next 50 years.[53] Increased knowledge of warming trends and associated effects would allow full economic accounting of costs of controlling emissions of greenhouse gases. Accounting will also include costs incurred from environmental damage that may occur with global warming and associated climate change. The costs of unabated greenhouse emissions, temperature increases, and associated effects are potentially huge because climate influences almost all aspects of our economic systems.

Sunburning Earth: Ozone Depletion

Ultraviolet (UV) radiation is high-energy light produced by the sun that damages DNA and can lead to mutations. This UV light is dangerous to humans because it can cause skin cancer, suppress the immune system, and damage eyes. It similarly harms all animals and plants. Even microbes are susceptible to damage by UV radiation, including microscopic plants in the ocean that are major consumers of atmospheric CO_2.

Ambient levels of UV light that have occurred naturally for billions of years cause a baseline level of health and environmental damage that is part of the cost of living in sunlight. Although most organisms (including humans and our food crops) have evolved to tolerate these natural levels of UV rays, some species (for example, those living at high altitude) have evolved to tolerate even more intense UV light. Still, many species are injured and some ultimately die if UV levels exceed historical levels. Interactions with UV light and pollutants can exacerbate UV sensitivity.

Ozone (O_3) high in the atmosphere protects all life on the surface of Earth from deadly exposure to UV rays. Ozone slowly forms from the oxygen (O_2) in air upon exposure to sunlight. This slow replenishment of the protective ozone in the stratosphere (at an altitude of 12 to 24 miles, or 20 to 40 km) absorbs harmful UV radiation from the sun and maintains the protection. In 1970, scientists expressed concern that gaseous releases of supersonic airplanes would deplete the protective layer of ozone in Earth's atmosphere faster than it is replenished. Then, by 1974, researchers had evidence that chlorine- and fluorine-containing organic compounds (CFCs) used in spray cans and air conditioners were entering the stratosphere and destroying the ozone. Because CFCs are nontoxic and do not react with many chemical compounds, they were used heavily in refrigeration, in spray cans, to make foam plastics, and as cleaning agents in the electronics industry. Technical details of how CFCs behave in the atmosphere are somewhat complex, but the crux of the matter is that they can move into the upper atmosphere, where each CFC molecule destroys many ozone molecules.[54]

By the 1980s, scientists documented that a large "hole" in the ozone was developing over Antarctica each spring. The hole is a thinning in the layer that has gotten progressively worse since 1980 and is growing larger (figure 3.4; the hole at its summer maximum currently exceeds the size of Antarctica). A similar, but smaller hole forms over the Arctic. This response to CFC pollution was dramatic and unpredicted. Who would have thought that simple aerosol cans, refrigerators, and air conditioners could have such drastic effects on our globe? Since the Antarctic ozone hole was discovered, additional decreases in ozone have occurred in the high atmosphere across the globe, but none as significant as those related to the Antarctic ozone hole.

Acknowledging worrisome damage to the ozone, signatory nations of the Montreal Protocol agreed in 1990 to completely stop emissions of CFC and related compounds by 2000. All the signatory nations did not meet that goal, however, and the ozone hole continued to worsen during the 1990s. Finally, in the early 1990s with agreements for financial compensation, China and India agreed to limit emissions of CFC compounds. Today, not all countries have limited emissions, and, to be sure, substitute compounds may be as bad as CFCs. Even if emissions were curtailed, it is not clear how long it would take for Earth's atmosphere to recover. And some unforeseen problems could worsen the damage—for example, increased use of hydrogen fuels to replace gasoline could lead to greater hydrogen releases into the atmosphere, and this would exacerbate the ozone-depletion problem.[55]

For more than 25 years, CFC release has been a recognized agent of global pollution that can lead to serious human health problems. People have been working to decrease emissions for more than two decades. Atmospheric scientists do not know how much worse the

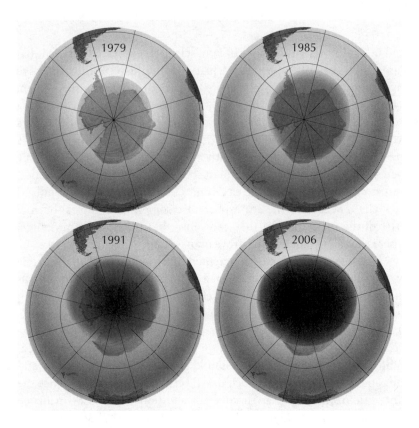

FIGURE 3.4 Increasing size of the ozone hole from 1979 to 2006, viewed from above the South Pole. In 1991, the Montreal Protocol was signed, and the size of the hole has increased substantially since that time. Note that the hole reached the southern tip of South America (Patagonia) by 2006. An increase in the intensity of ultraviolet radiation related to ozone depletion has occurred in inhabited parts of the world, but dangerous increases in ultraviolet rays are most notable in the southern portion of the Southern Hemisphere (Chile, New Zealand [R. McKenzie, G. Bodeker, and B. Connor, "Increased UV Radiation in New Zealand: A Cautionary Tale," *NIWA Water and Atmosphere* 7 (1999): 7–8], and Australia). People in this region are experiencing several times greater exposure to ultraviolet radiation than people in places that are not under the ozone hole, and they are regularly advised to protect their eyes and skin from demonstrably higher ultraviolet exposure. (Images courtesy of NASA)

ozone problem will get or whether current attempts will correct the problem, and if they do, exactly how long it will take to reverse itself. An analysis by NASA scientists suggests that the ozone is recovering as a result of CFC controls, and press releases indicate the ozone should be back to pre-1980 levels in the next 50 years.[56] The short period of time it took to damage the ozone, the time lag until political agreements could be implemented to control CFC releases, and the long period of time required for ozone recovery are other examples of both political and natural momentum of a global environmental problem. But there is cause for optimism: at least humanity made a major attempt to correct a global environmental problem, and the ozone hole may eventually close.

Agriculture Is Fouling Our Water

Humans have expanded the amount of Earth's surface used for agriculture and modernized the methods used to grow crops. This increased cultivation has resulted in the acceleration of water pollution on a global scale: sedimentation (choking waters with soil from erosion) and runoff of fertilizers and pesticides are major problems associated with agricultural practices.[57] In the United States, assessment of water quality indicates that agriculture is the single largest cause of degradation of stream and lake water quality.[58] Other parts of the world have the same problems, except those areas that are so heavily urbanized or industrialized that there is little or no agriculture at all. Such urbanization and industrialization also degrades water quality, of course.

Fertilizers

Over the past 30 years, global rates of fertilizer use have expanded drastically (figure 3.5). Today, the rate of nitrogen production from agriculture is about equal to that produced naturally, and the amount of phosphorus removed from the ground by mining is roughly the same as natural rates of phosphorus release from rocks and soils by weathering (for example, leaching by rain).[59] Approximately 6 ounces (170 g) of nitrogen fertilizer are used to produce each pound of food we eat; of this, 90 percent leaks into the atmosphere, soil, and water.[60] As agricultural food production causes fertilizers to enter our water,

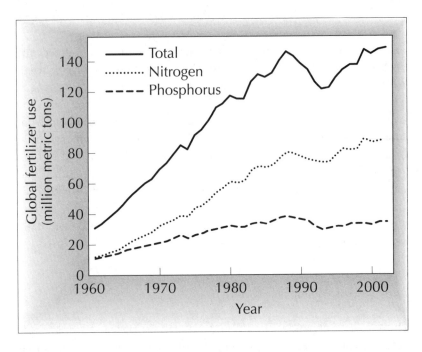

FIGURE 3.5 World fertilizer use from 1960 to 2004. The current use of fertilizer is double the natural rate. Much of the fertilizer added to cropland is not taken up by plants but washes into aquatic environments, causing nutrient pollution problems.

concentrations of nitrogen and phosphorus increase 50 to 100 times or more over historical baseline values of streams and rivers draining agricultural areas.[61] Increased nutrient concentrations are evident in every part of the world with significant agricultural activity, and this form of water pollution can cause human health problems, degrade water quality, and harm ecosystems.

Nitrogen can be toxic to humans in high concentrations. Elevated nitrogen intake (in the form of nitrate, a common component of fertilizer) can cause a condition called methemoglobinemia (try saying that three times fast!). Nitrate can be converted to toxic nitrite in the stomach. In this syndrome, nitrite attaches to hemoglobin, making the hemoglobin defective and unable to carry oxygen; without oxygen transport to cells, humans suffocate. When it is not bound to oxygen, hemoglobin is blue. Infants are particularly susceptible to this form of nitrogen poisoning and turn blue as oxygen is displaced from the hemoglobin in their blood. Thus this condition has been given the non-technical name "blue-baby syndrome." People of all ages will die if they ingest enough nitrate; those who are elderly, very young, or ill are most susceptible.

Some additional negative effects of global water pollution caused by fertilizer inputs are found in all types of freshwater around the world.[62] Fertilizers stimulate the growth of huge populations (called "blooms") of microscopic plants—in particular, noxious algae. Normal amounts of algae are good, but excessive algal growth can ultimately reduce the amounts of oxygen that are dissolved in water, harming animals that live there. In some cases, massive fish kills occur because dissolved oxygen is removed from water, resulting in low oxygen conditions. Excessive algal growth can also cause taste and odor problems in lakes. Many cities that rely on surface water for drinking are plagued by bad tastes related to blooms of algae that are stimulated by runoff of agricultural fertilizers. Dallas, Wichita, Phoenix, Tacoma, and Chicago are just a few of the cities in the United States and around the world that historically have been plagued by algal blooms in their drinking-water sources.

Some algal species that bloom in response to fertilization are toxic and, if ingested, can cause the death of humans and other animals.[63] Other species of algae synthesize small amounts of toxins that cause long-term human health problems, such as liver and neural damage. These algae can grow in lakes used for drinking water, and most conventional water-treatment processes do not remove many of the toxins. Drinking-water regulations to control concentrations of toxic chemicals are currently being instituted in many developed countries, including Australia and Canada. The United States has yet to enact nationwide regulations on algal toxins in drinking water.

In coastal marine systems, algal blooms stimulated by nutrient pollution can cause toxic effects such as "red tide." Blooms of some species of red tide algae can become very large and cause large-scale die-off of marine organisms, resulting in massive economic costs.[64] Other algal species in estuaries in the northeastern United States have been reported to cause substantial health problems.[65] These toxic algae excrete potent neurotoxins that can harm people who consume or work around water containing them. Beaches are closed, fishery workers must stay off the water, and many fish die when this type of algae blooms in estuaries with nutrient pollution. Other species of toxic algae may contribute to seal deaths in Europe and elsewhere.[66] Harmful algal blooms occur in coastal regions around the world.[67]

Nutrient-stimulated blooms of toxic algae can increase the probability of cases of human poisoning from seafood consumption. Shellfish filter algae from water and can concentrate toxins in their body tissues, causing severe illness or death of humans who consume them. Large fish on some coral reefs eat other fish that eat invertebrates that consume toxic algae. Toxin is concentrated within each organism in the food chain, with concentrations increasing from invertebrate to small invertebrate-eating fish to large fish that eat small fish.

Stimulation of algal growth can be harmful, even if algae that are stimulated by fertilizers are not toxic. Coral reefs are endangered worldwide, and nutrient enrichment is thought to contribute in part to their decline.[68] Fast-growing algae stimulated by fertilizer runoff may smother coral reefs, which are important to maintaining CO_2 levels in the atmosphere.

Coastal areas can be harmed by fertilizer runoff that stimulates algal growth and causes reduced oxygen. The Gulf of Mexico has experienced a low-oxygen condition over many hundreds of square miles (called the "Dead Zone") that has harmed substantial numbers of fishes and shellfishes, probably damaging the fishing economy. This low-oxygen condition has been directly linked to fertilizers from agriculture in the midwestern United States entering the Gulf via the Mississippi River and stimulating algal growth.[69] The problem has increased in severity over decades, and warming associated with the greenhouse effect is probably making it worse because warmer water temperatures make it more difficult for water to mix.[70] Similar "dead zones" form in many parts of the world where agricultural fertilizers enter gulfs or other marine areas that have restricted water exchange with open ocean. This fertilization endangers sea grass beds, kelp beds, and coral reefs.[71] Beds and reefs are nurseries for many economically important fishes.

Agricultural fertilization is global now, and, consequently, unwanted stimulation of algae by agricultural nutrients is occurring throughout the world. Essentially all marine and freshwater habitats that receive runoff from land used by humans are afflicted by this problem. Different areas exhibit various degrees of severity of nutrient pollution, with the greatest problems occurring in areas with the most intensive agriculture.

We do not know how long it would take to reverse the effects of decades of increased fertilizer inputs into the environment if we were to drastically curtail runoff of agricultural fertilizers. Aquatic communities naturally recycle nutrients; organisms retain and reuse nutrients very efficiently. Calculations for some lakes indicate that it would take centuries for the effects of excessive phosphorus fertilization to be mitigated naturally.[72] Much fertilizer added to farm fields cannot be accounted for and, thus, will eventually leave fields and likely end up in rivers and streams.

Nutrients and water quality are my specialized area of scientific research. I believe that even if all fertilizer use were to cease immediately, it would take at least half a century for nutrient pollution to leave our watersheds. This length of time to recovery is related to the fact that organisms in streams and other habitats retain and release nutrients very slowly.[73] Furthermore, once pollution contaminates groundwater, it can take decades or even centuries for water to flow through the ground and enter lakes, rivers, or oceans. Before the harmful effects of nutrient pollution in our waters can be abated, nutrient inputs will have to be controlled. But because we must grow food for more people on less land, fertilizer use is not likely to decrease substantially in the near future.

Pesticides

Pesticide use also is continually increasing globally (figure 3.6), with billions of pounds used each year. Chemical pesticides include herbicides (to control unwanted weeds), fungicides (to control fungal infections), and insecticides (to control invertebrate pests). Although it is desirable to produce pesticides that target only the species of interest, it is often not practical to do so. Pesticides work because they are toxic. All living organisms share many of the same basic biochemical pathways, making it difficult to find compounds that are toxic to only one very specific organism. As we intensify agricultural production, more chemicals are required. Importantly, while many individuals of the target species die, a few survive and are almost certain to evolve with tolerance to the chemical. Once tolerant pests take over, even more pesticide is required to control them. Even with use of pesticides, about 40 percent of food production is still lost to insects, weeds, and plant diseases, and then pesticides are added again.

Pesticide use without repercussions in our water quality is not possible. For example, annual costs associated with the use of pesticides in the United States alone are estimated to include $1.8 billion for cleaning

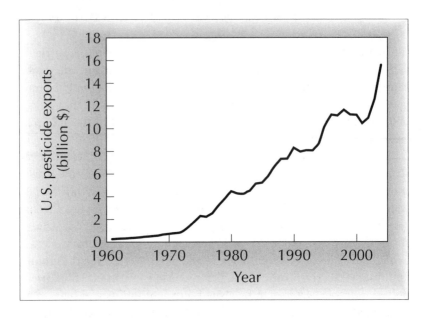

FIGURE 3.6 Value of pesticides exported by the United States from 1962 to 2004. Export rate is a good indicator of global pesticide use rates, since the United States is a major producer. Crop losses to pests have not declined proportionally to increases in global pesticide use rates.

groundwater, $24 million in fishery losses, and $2.1 billion in losses of terrestrial and aquatic birds.[74] Loss of human health and lives costs even more.

In developed countries, pesticide use is regulated to minimize the negative effects on human health, but in many developing countries, regulation is absent or poorly enforced. Chemicals are being distributed to people around the globe who often do not have proper equipment or training to use them safely.

Many pesticides persist in the environment for decades. The insecticide DDT (dichloro-diphenyl-trichloroethane) provides an example of how persistent these compounds can be. DDT was used heavily in the United States in the 1950s and 1960s, until scientists realized it was being concentrated in animals in the environment. Biological concentration of DDT almost led to the extinction of peregrine falcons and bald eagles due to the process of biomagnification within the food chain: DDT was absorbed by microscopic plants, concentrated by small crustaceans that ate plants, further concentrated by small fish that ate crustaceans, and again by large fish that ate small fish. Concentrations in the bodies of eagles that ate fish (or, for that matter, humans who ate fish) could be millions of times greater than concentrations in water where DDT was originally found.

The use of DDT was banned in the United States in 1972. As of 1998, only three countries allowed DDT production: India, China, and Mexico. But chemical companies still stockpile DDT. Although DDT can be broken down slowly by microbes, it persists in many environments. In the United States, DDT was used for about 30 years, but it remains in the environment almost 40 years after its use was drastically decreased.

Not only has DDT been transported around the world, but resistance to it has been transported to pest insects that it targets around the world. The gene for DDT resistance in fruit flies has been found in fly populations across the globe.[75] Thus DDT residues and the ability of pests to survive exposure to DDT have spread globally. If we stop using DDT, levels in the environment will eventually become undetectable over the coming decades; nevertheless, the genes in insect pests for resisting DDT will likely remain for much longer.

Global Spread of Toxins

We produce many organic chemical toxins, including pesticides, which are released into the environment and spread around the globe. Persistent organochlorine compounds (compounds containing carbon, oxygen, hydrogen, nitrogen, and chlorine) are found worldwide.[76] Many of these compounds are long lived and dispersed through the atmosphere, even to remote habitats. Compounds condense from the atmosphere, depending on temperature, with the most volatile organic chemicals condensing in either high-altitude or polar regions.[77] Tissue samples from Arctic seals contain high concentrations of stain repellents (for example, Scotchgard).[78] Polar bears have increasing concentrations of mercury, more than 14 times greater than were present in samples from 700 years ago.[79] Heavy contamination is common across the Russian Arctic, with high concentrations of the insecticide HCH (hexachlorocyclohexane) in the breast milk of indigenous women.[80]

Atmospheric transport moves compounds long distances, concentrating them far from their point of origin. Thus we can find organic contaminants in the bodies of albatrosses (an open-ocean seabird) in the middle of the Pacific Ocean, far from locations where these chemicals are synthesized and used by humans.[81] Roger Payne, a professor who discovered that humpback whales sing songs, is currently sailing around the world, taking small tissue samples from whales. These samples are sent back to laboratories for analysis. All whales, even those in the most remote oceans, are contaminated with organic pollutants released by humans on land.[82] Remote glaciers contain contaminated ice from precipitation over the past few decades, and these contaminants are being released into the ocean as global warming causes glaciers to shrink.[83] This glacial melting, and the long lifespan of whales, explains why Payne's samples from whales often contain DDT, which is not currently used heavily but is quite persistent.

A survey of atmospheric samples for pesticides in North America revealed substantial contamination of the air we breathe. In 2000 and 2001, some 40 stations from Central America to Canada were sampled for airborne pesticides. DDT was found in Mexico and Belize, indicating ongoing use there. The new generation of pesticides used in the United States and Canada was more common in samples from those regions. All these compounds are meant to be added only to crops or vegetation, but they were commonly transported long distances in the atmosphere.[84]

Transport of pesticides from the Central Valley of California into the Sierra Nevada is substantial. These pesticides have reached lethal levels in some species of frogs and fish. Thus anglers in the apparently clean Sierra Nevada could consume substantial amounts of agricultural pesticides in fish they catch, even though they fish far upstream (but also upwind) of any agricultural activity.

The amount of wind transport from agriculture and declines in frog populations were clearly associated for the endangered California red-legged frog. Researchers compared measurements of the amount of total surrounding nearby agricultural land with the amount of upwind agriculture; they found that wind direction, and thus atmospheric transport from fields, was a factor in frog-population decreases.[85] Further studies indicate that pesticides harm at least four other species.[86] Pesticide toxicity, coupled with diseases (chapter 4), predation, and competition from bullfrogs may ultimately spell the demise of the red-legged frog.

An extreme example of transport of agricultural compounds has been documented in subarctic North America. Atmospheric transport, in combination with biomagnification of a long food chain (as described for DDT), can account for unusually high concentrations of the crop pesticide toxaphene in fish collected from a subarctic lake.[87] One would assume that this lake is very clean because it is far from civilization and any crop production. The fact that toxaphene heavily contaminates fish in this lake illustrates the global nature of human chemical use, that toxic organic materials persist in the environment, and that chemicals move far from their original point of use.

Nobody can predict with certainty the interactions of the toxic organic compounds we spread around the globe. These compounds interact with humans in unpredictable ways, sometimes in combination. A survey of 139 impacted streams from 30 states in the United States, conducted in 1999 and 2000, revealed that 80 percent of streams had detectable levels of at least one of 95 organic chemical compounds assayed. The typical stream had seven compounds, and some streams had as many as 38. Chemicals ranged from pharmaceuticals and hormones to insecticides. The most common compounds were steroids naturally produced by humans and other organisms, insect repellant (DEET [meta-N,N-diethyltoluamide]), caffeine, fire retardants, disinfectants, and detergent compounds.[88] We know almost nothing about how these chemicals combine to influence humans and wildlife.

Hormones and Antibiotics

One unpredicted mode of toxicity of organic chemicals discovered only recently is that they mimic and disrupt natural hormones. Only in the 1990s, some 40 years after DDT use began, did it become a concern that many toxins such as DDT act as hormone mimics and disrupt natural biological processes. Numerous organic compounds can mimic natural chemicals found in our bodies and those of other animals. These natural chemicals, called endocrines, regulate the way animals function by controlling the chemical balance in the body. Toxic compounds that mimic these endocrines are said to lead to endocrine disruption.[89]

Tyrone Hayes is a respected professor at the University of California at Berkeley and a researcher who published some of the first studies on the effects of low levels of agricultural chemicals on amphibian health. He found that vanishingly small concentrations of atrazine, the most common agricultural weed killer, caused feminization of leopard frogs in the laboratory and found similar effects in wild populations.[90]

In 2003 alone, 77 million pounds (34 million kg) of atrazine were applied in the United States. One part in a billion of atrazine inhibited larynx growth in African clawed frogs. Hayes's research was suppressed and criticized by the chemical industry.[91] Political battles over this compound illustrate the high economic stakes of agricultural chemical industries and how hard they work to continue their chemical sales. The actual research on atrazine demonstrates how very small concentrations of widely used chemicals can have unintended and profound effects on animals.

A number of chemical compounds are endocrine-disrupting because they mimic estrogen. These compounds include DDT, which has recently been linked to nonfunctional testes in male alligators and to other reports of feminized wildlife.[92] Constant exposure to chemicals that mimic estrogen will also have negative impacts on human health. For example, some investigators claim that recent substantial decreases in human sperm counts in developed countries are related to exposure to hormone mimics, but this remains to be proved.[93]

It took many years of research to recognize that organic chemicals released into the environment can mimic hormones and, if timing of exposure is just right, cause devastating effects on the offspring of animals. The potential for inadvertent biological activity that decreases intelligence, fertility, and resistance to disease is poorly documented, even though most humans are exposed to multiple synthetic organic chemicals every day. Researchers can barely keep up with testing the effects of numerous new pesticides on the market, never mind the interactions of multiple compounds with the complexity of human physiology.[94]

Another poorly recognized global effect of the use of agricultural chemicals is the widespread addition of antibiotics to animal feed. These antibiotics increase the growth of livestock, but they also select for antibiotic-resistant bacteria that are spread into rivers and from there around the biosphere. It is virtually certain that bacteria will evolve with the ability to resist any new antibiotic. Each antibiotic humans have synthesized has led to the evolution of resistant strains of bacteria. These bacteria contain genes for antibiotic resistance that can be spread to other bacteria.

The inevitable spread of antibiotic resistance from harmless bacteria to types of bacteria that cause human diseases has been well documented. In just one example, the antibiotic nourseothricin was used in Germany in the pig-breeding industry starting in 1980. Within two years, antibiotic-resistant bacteria were found in pigs, manure, river water, food, and humans.[95] First, antibiotic-resistant bacteria were found in people who worked with the pigs, then in their families, and finally in the entire population of a town near the pig farms. After that, resistance spread to bacteria that infect humans; it was isolated from bacteria that had infected local hospital patients. Even after nourseothricin use was discontinued, resistance remained for years in types of bacteria that cause human disease.

A serious problem with agricultural antibiotic use is the inevitable evolution of "superbugs" that cause disease in humans and are resistant to drugs that we currently have to control them. These bacteria are able to combine genes coding for resistance to multiple antibiotics into circular pieces of DNA that can be passed among strains of bacteria. Multiple antibiotic resistance is so ubiquitous that it has spread throughout the world and can be found in bacteria in the middle of the ocean and in other areas far removed from agricultural, industrial, or medical areas where the genes arose. Genes for antibiotic resistance will continue to spread into disease-causing bacteria, and once there, they are not likely to leave. The continued use of antibiotics in agriculture exacerbates the problem.

The net effect of our use of chemicals for agriculture and other purposes is that we are loading all our natural ecosystems with detectable levels of toxins. This will almost certainly continue as we attempt to feed ever more people and provide more people with a meat-rich diet. More meat means more crops, which require more pesticides. Pressure will mount to use even more agricultural chemicals to protect crop yields as demand for food increases.[96] Thus we can predict, based on past trends, that pollution and habitat destruction associated with agriculture will escalate in the next 50 years. To feed the number of people who will inhabit Earth 50 years from now will require conversion of an area of land larger than the surface area of the United States from natural habitat to agriculture, and this will also lead to 2.4 to 2.7 times greater rates of fertilizer and pesticide use. These extensive changes are bound to aggravate problems associated with chemical use for agriculture, in addition to those discussed in the section on species extinctions.[97]

Chemical and physical perturbations of Earth's biosphere are only part of the global pressure exerted by humans, and this chapter describes only the barest details of those perturbations. The shock wave of these disturbances alters the environment of all plants and animals (including ourselves). Our footprint also directly influences many of the species on Earth by introducing unwanted species and causing the extinction of many others.

Humans must learn what are nature's dynamic
capacities, because excessive violation without harmony
will only unleash her intolerable vengeance.

—Robert G. Wetzel

Weeds and Shrinking Violets

Pests on the Move and the Ecological Holocaust

In many parts of the world, urban areas are almost completely devoid of native plants; gardens, whether commercial or personal, are planted with species bred by humans for show or hardiness under polluted conditions. On the Hawaiian Islands, which once contained numerous unique species that evolved there, human-caused species introductions and habitat destruction have made it so almost all plants and animals there now are nonnative. All other areas inhabited by humans are experiencing similar issues to varying degrees. Today, native species around the world are either endangered or already extinct. We are close to losing some of the iconic species of our planet, including tigers, rhinos, and several species of whales. We are also losing countless other less-charismatic species.

Another global consequence of the shock wave of the insidious explosions of population growth and resource use rates is collateral damage to other species. As species are introduced at greater rates than ever before, other species are extinguished. Habitat destruction and degradation (toxins and pollutants), direct removal of some species, and introductions of others are now causing one of the greatest extinction events in the 4.5-billion-year history of Earth. Shifts in abundance and distribution of other species on Earth are immense and have grave consequences. Humans, however, may slow or halt global environmental change before momentum takes us to the point where nature's "intolerable vengeance" is wrought.

Visitors Worse Than In-Laws: Weedy Species Never Leave

Invasion by nonnative species is a ubiquitous, global, costly, but often overlooked effect of humans on the environment. People now move more materials more rapidly across greater distances than ever before, and this greater speed and distance of transport increases the probability of unintentional species introductions. Furthermore, species are being introduced intentionally into many habitats: introduction of pests can have strong economic consequences, in addition to the negative environmental effects. However, many purposeful introductions of species thought to be useful have resulted in the establishment of pests.

Unlike easily accounted for increases in carbon dioxide, decreases in ozone, and fertilizer production and use, no exact estimates of total global rates of harmful species introductions are available. Numerous unwanted, introduced species have caused enough of a nuisance and economic loss to be noticed by humans; some case studies are discussed in this chapter, but more inconspicuous introductions are not documented. For example, if a nonnative insect is introduced, it may not be noticed unless it causes crop damage, even if it has strong effects on parts of the ecosystem that are of concern to humans.

Many potential negative economic effects are caused by the introduction of unwanted species, including (1) loss of valuable biodiversity (in the United States, 49 percent of endangered or threatened species are under pressure from invasive species),[1] (2) loss of beneficial ecosystem functions (for example, water purification), (3) destruction of preferred species, including agricultural organisms, and (4) introduction of human diseases and the organisms that carry them.

One disease that was introduced along with nonnative species is bubonic plague, or the Black Death. The bacterium that causes the plague is carried by ship rats and transmitted by fleas that bite the rats and then bite native rodents. Over the past century, bubonic plague has spread into ground squirrels, prairie dogs, chipmunks, mice, and wood rats in the western United States. West Nile virus is transmitted by mosquitoes but can find a reservoir in birds or horses. Although West Nile virus was first described in humans in the 1930s, it did not enter the United States until the 1990s, possibly in an infected bird introduced from overseas. An African species of mosquito brought dengue and yellow fever to South America, causing hundreds of thousands of deaths. An African tick introduced into the Caribbean in the 1960s carried several human diseases, including rickettsia. Movement of poultry and other birds is causing the current spread of avian flu. Human health has been compromised repeatedly by introduction of animals around the world.

Once an unwanted species becomes established, it is almost impossible to eradicate. The science of species introductions is most often approached as a series of case studies because of the difficulty of predicting which species will become established and which will have negative effects if they do. Some species have tremendous negative impacts.

One example of an introduced species that can inflict serious harm on native animals is the Nile perch in Lake Victoria. Nile perch were introduced in 1957 to this huge lake, which straddles Uganda, Kenya, and Tanzania. Within 20 years, this large predatory fish helped eliminate more than half the lake's 200 fish species. Fish species in Lake Victoria ranged from those prized by aquarists to those that provided badly needed protein to the local human population. Most lost species were unique; they evolved in the lake and do not occur anywhere else in the world. Nile perch is so large that it is difficult to catch using traditional methods and has decimated the traditional fishery, causing economic hardship for many people who inhabit the lakeshore.[2] The perch is taken commercially by larger nontraditional boats and methods, but it is too expensive for most local people to afford. Nile perch will be extremely difficult, if not impossible, to eradicate. Despite the positive economic effect of more exported fish, overall negative costs have been tremendous and include the irreversible loss of numerous fish species.

Zebra mussels, small, clam-like animals, have invaded North American freshwater lakes and rivers. They were introduced into the Great Lakes in the late 1980s from ballasts of ships from Europe. By 2000, they had spread over a large portion of the Mississippi River drainage. These mollusks clog intakes of drinking-water treatment and industrial facilities, causing millions of dollars of damage.[3] They outcompete native mussels and severely alter natural food webs (the complex feeding net of natural animal communities).

A strange twist in the zebra mussel story is that pollution control in Europe may be responsible for their spread to the United States. Zebra mussels came to the United States from western Europe. Transatlantic

65

ships had been taking on ballast water from European freshwater ports and releasing it in the Great Lakes. These ports in Europe were too polluted to allow the survival of the mussel until pollution controls were enacted, and then live zebra mussels or their larvae began finding their way into ballast water. Ships then crossed the Atlantic Ocean, dropped ballast water into the Great Lakes before taking on cargo, and in the process released the noxious invader.

The invasion of zebra mussels in western Europe as well as the United States has been aided by canals constructed to facilitate barge traffic. Even before the mussels entered the United States from western Europe, they invaded western Europe from Baltic areas when barge-transport canals were constructed across drainages that had not been connected previously. In the United States, once the invader reached the Great Lakes, it spread through shipping canals that link the Great Lakes to the Mississippi basin. From there, it has rapidly spread across the central United States (figure 4.1).

Zebra mussels have an expansionary history because each female can produce more than 1 million offspring each year. Small zebra mussel larvae can be easily transported and come to dominate new habitats. Adults can attach to the bottom of boats and survive for several days when a boat is moved from one body of water to another. These characteristics optimize the chances for spreading by human activities. The most optimistic scientists think that we will discover a method to reduce the number of zebra mussels, but none claim that we will ever eradicate them from waters they have invaded.

In 1859, some 24 wild European rabbits were introduced into Victoria, Australia. There are now more than 200 million throughout the country, where they cause major damage to farmland and wild areas. An estimated $50 million is lost in livestock feed, and $10 million is spent on control of rabbits each year in Australia. Rabbits directly led to the extinction of two species of burrowing marsupials by consuming forage these animals needed to survive. In addition, rabbits allow more predators to survive and thus increase predation pressure on rare animals. Rabbits eat native plants and greatly accelerate erosion rates, causing loss of valuable topsoil and pollution of aquatic habitats by sediments. Introduced rabbits have had a large influence on ecosystems across Australia.

A virus was introduced to control the Australian rabbits in the mid-twentieth century and remained effective for a time. But as the virus became less deadly (it evolved to not kill its host before it could

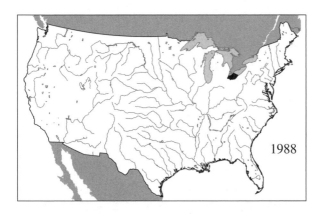

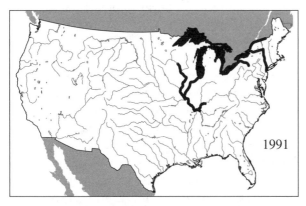

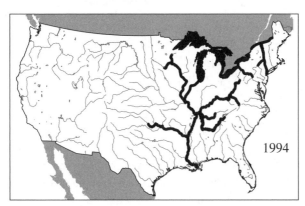

FIGURE 4.1 Explosive increase in the distribution of zebra mussels in the United States from 1988 to 1994. The dark line indicates rivers and lakes where the mussels have been recorded. Zebra mussels had spread into five more states by 2005 and will eventually be found across the continent. Boat hulls with zebra mussels have been found in 11 other states, but the mollusks have not become established there yet. (Data courtesy of the U.S. Geological Survey)

spread), rabbits evolved some resistance to infection by the virus and the rabbit population exploded again. A second virus has been released, and many rabbits have died, leading to improvements for native plants and animals. The second rabbit decline is probably only a temporary victory (or setback, if you are a rabbit), as earlier experience demonstrated. We know that evolution will lead to disease resistance and decreased virulence of the virus. Rabbits will develop resistance to the new virus, and strains of the virus that are too deadly will die out with their victims before they can be transmitted to other rabbits. Fairly soon, the rabbit populations will once again expand. European rabbits now appear to be permanent residents of Australia.

Unwanted weedy plants also cause serious problems around the world. For example, leafy spurge is a major nuisance in the western United States.[4] Leafy spurge, an herb, is a native of central and eastern Europe. It was probably introduced into the United States in the early twentieth century from imported oats. Although spurge is not eaten by livestock, it outcompetes desirable vegetation, and it is estimated to be responsible for direct losses of $58 million in the upper Midwest.[5] This economic estimate includes only severe degradation of range quality, not the cost of reduced biodiversity when spurge overruns native plants and decreases forage available for wildlife. Leafy spurge cannot be controlled easily.

Numerous other plants cause ecological and economic problems in the United States, including water hyacinth, water milfoil, purple loosestrife, kudzu, spotted knapweed, gorse, Scotch broom, tamarisk, and cheatgrass. Introduced nonnative bees may facilitate the establishment and spread of these unwanted plants.[6]

Islands are particularly vulnerable to invasions because their inhabitants have not evolved to be highly competitive or resistant to predation. The brown tree snake in Guam illustrates the massive devastation that can be caused by island invaders.[7] The snakes became established in Guam in the 1950s, and they are now so numerous (20 per acre of jungle, or more than 50 per ha) that they often enter power transformers and lines and are responsible for numerous power outages each year. Brown tree snakes climb trees easily and are effective predators on bird nests. The snake has wiped out most native birds and many introduced birds. While economic costs are high, a forest without birds is not complete. Not only is the beauty of their song and plumage missing, but their ability to control insects and to pollinate and disperse fruits (propagate plants) has been lost. The snakes have built large populations that now must subsist on lizards and other smaller reptiles, which, as a result, are now endangered. Since the snake is not native to Guam, it has no indigenous predators. Brown tree snakes have been found in Hawaii at least six times, in Rota in the Mariana Islands, and on the mainland United States, but luckily these snakes have been prevented from establishing populations in these places.

Scientists have a tough time predicting which species will become successful invaders.[8] Only a minority of introduced species become established, and of those that do, many take a small place in the environment and cause no damage. Nevertheless, given the tremendous damage that can be caused by an individual invader, and the difficulty of predicting which invader will become a pest, it is advisable to try to prevent the introduction of any species into new habitats.

Human activities can allow the transportation of organisms in unusual and unpredictable ways. For example, plastic debris in oceans has become common worldwide. One investigator characterized debris on 30 remote islands, from the Arctic to the Antarctic. He found that 20 to 98 percent of debris was manmade (yet another type of global pollution). Many living organisms were attached to debris, including some with no other modes of long-range dispersal. The researcher estimates that human rubbish has more than doubled rates of movement of animals attached to floating materials in the subtropics and more than tripled rates in temperate and arctic regions.[9]

Wherever introduced pests have become established, environmental damage gains an extreme amount of momentum. All that is necessary to start an invasion is one sexually reproducing pair, or one individual if reproduction is asexual. In most instances, once established, the pest will never be eradicated. As the amount of materials transported around the globe expands exponentially, so will the number of unwanted species introductions. Increased movement of viable pests is attributable not only to the number of people and amount of materials moving but also to travel being so rapid that pests are more likely to survive the journey. We are moving forward into a world with more and more "weedy" species that are difficult, if not impossible, to control. As human activities cause extinctions of sensitive species, weedy invaders increase.

Species Extinctions

Another clear global effect of humans is a tragic decline in species. More than one-half of the species on Earth could be extinct in the next 100 years. Humans are causing one of the few major extinction events that have occurred over the past billion years.[10] We are living in an ecological holocaust. Since 1950, upper-limit estimates are that 300,000 species on Earth have gone extinct,[11] and many other species that were closely associated with those species may have gone extinct as well.[12] One-third or more of amphibians are threatened with extinction.[13] Around 10,000 species of freshwater invertebrates are imperiled or extinct.[14] More localized losses of plants and animals are common; about 13 percent of butterflies in Britain have been extirpated (locally exterminated).[15] Extinction is forever, and humans are executioners in this current extinction spasm.

The primary cause of species extinction is habitat destruction and loss. Human impacts such as agriculture, urbanization, logging, mining, and other development all take their toll. For example, biodiversity hot spots (areas with many species) are threatened because people are living in bigger houses. Where there are more households per person, and suburban sprawl eats up wild habitat, species are under increasing pressure.[16]

Bird, Mammal, Fish, and Frog Extinctions

For some animals, such as birds and mammals, most species are described, and we have a good idea of why and when they become extinct. The extinction rate of birds has soared over the past 200 years, and this rate is consistent with increases in human population, resource use rates, and habitat destruction (figure 4.2). Loss of bird species is worrisome because birds are good bioindicators. They are global "canaries in the mine" that can be used to signal problems with other species. Birds require specific types of habitats, are well studied, and are conspicuous. When a bird species has not been observed for some time in the wild, it is likely extinct. Occasionally, a species thought to be extinct, such as the ivory-billed woodpecker, is reported found. But, unfortunately, finding a species after decades is the exception. Thus we can be confident in reported extinction rates of birds. Extinction of birds is also a likely indicator of extinction of lesser-known species. If the habitat of a

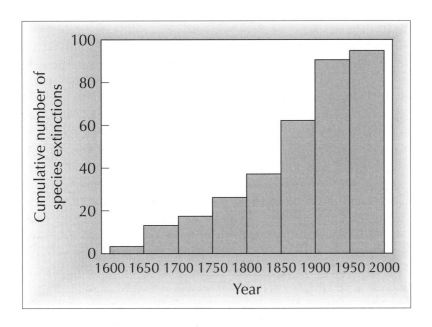

FIGURE 4.2 Escalation of bird extinctions from 1600 to 2000. Each bar represents the total sum of the number (cumulative) of documented extinctions since 1600. (Data from G. Nilsson, *The Endangered Species Handbook* [Washington, D.C.: Animal Welfare Institute, 1983])

bird is destroyed to the point that it has been forced to extinction, the probability is high that many other less-studied species have also been lost.

Not just birds are going extinct. Many species of mammals, fish, reptiles, amphibians, and plants are extinct, extirpated, or endangered around the world. For example, about one-quarter of terrestrial mammals are at risk of extinction.[17] About one-third of freshwater fishes are endangered. Virtually every well-studied group of animals is under significant pressure from human activities. Lesser-known species are also being lost.

Many species have positive effects, such as maintaining ecosystems that support us and as a potential source of novel medicinal compounds. We could be destroying the next cure for cancer to accommodate yet another shopping mall. Earth has been compared to a spaceship, with rivets that hold the spaceship together being equated to species on Earth. Each lost rivet has little or no impact on the structural integrity of the ship. When many rivets have been lost, eventually one popped rivet will cause catastrophic structural problems. Likewise, species that make up our ecosystems are all part of the structure that sustains us. We can lose some species without much damage. We do not know exactly which species removals will cause catastrophic effects because species interact in complex ways. Deleting species from Earth is a form of gambling with our biosphere or, to use the earlier simile, like poking a monster with a stick. Even if we know for certain that a species is an "unimportant rivet," it is immoral to cause a species to disappear forever.

Frogs are in decline or extinct worldwide in part from the movement of a fungal disease. The disease may cause the single largest extinction in a group of animals since the dinosaurs. Human activities likely caused the initial spread of the fungal disease. The African clawed frog could be a vector that carries the disease because it has been transported around the world for scientific research and a type of early pregnancy test, and it both carries the disease and is resistant to it.

Panamanian jungle streams, where the disease has not yet hit, are thick with tadpoles. Adult frogs are abundant in the surrounding jungle; in some places, numerous small frogs jump away in all directions with every step you take. Walking through the luxuriant vegetation near streams, you are surrounded by the sound of thousands of calling frogs. These beautiful creatures, including the bright yellow and black golden frog of Panama, will all be gone in a few years because few, if any, of the amphibians in this high-elevation forest are resistant to the disease. As the fungal disease sweeps in, first adults die and soon after tadpoles disappear.

When my friend Edgardo Griffith, a Panamanian frog researcher, found the first dead golden frog in the region most famous for these magnificent animals, he went home and cried. He knew first hand the devastation that was soon to follow. In regions of the jungle where the disease has swept through, dead frogs litter the forest floor within a matter of weeks. After the fungus moves through, the Panamanian jungle will be silent and changed forever. As frogs are stricken by the disease, they literally fall from the trees. Species of some of these tree frogs have no name because they have never been described by scientists. They will never be seen alive by humans.

Nothing can be done to stop the disease except to collect healthy frogs and put them into captivity, where they will not be exposed to the disease. We do know how to keep many of these frogs in captivity. Frogs are an important part of the ecosystems they inhabit. In high tropical jungles, tadpoles are essential to streams. They eat material that falls in and keep algae from choking the stream bottom. Adult frogs are food for the abundant snakes and lizards of the jungle.[18] Extinction awaits most susceptible frog species worldwide. In just one small group, the Panamanian golden frogs, around 30 of the 113 species are extinct and only 10 populations are not declining.[19] The effects of extinctions of many species of frogs will ripple through the ecosystems they live in.

In addition to movement by humans, global warming has accelerated infection rates caused by the fungus.[20] As species-rich, high-altitude rainforests warm, conditions become more favorable for the fungal disease to reproduce and spread. The double whammy of disease and global warming is ensuring that at least 60 species of frogs will die out in Central America. The disease has spread around the world and is endangering frogs from California to Southeast Asia. We may be the last humans ever to see high diversity of frogs in high-elevation tropical jungles and many other areas of the wild.

Harm to Our Ecosystems and Other Reasons to Conserve Species

The collapse of coastal marine ecosystems from fishing-induced extirpations is an example of how locally targeted species extinctions can have long-lasting effects on ecosystems, and such effects can occur over time scales that make it difficult for humans to perceive the severity of changes.[21] Historically, the numbers of large fish, sharks, turtles, and whales were tremendous in comparison with what they are now. For example, numerous islands throughout the oceans are named after species that are no longer there because they have been fished to extinction or at least to extirpation. Paleontological, archaeological, historical, and research records document numerous broad declines and extinctions of large marine animals. These declines, in turn, have altered community structure and endangered entire ecosystems.

Kelp (seaweed) forests of the Pacific Northwest provide an example of predator removal with cascading effects (figure 4.3). Sea otters were hunted to near extinction for their pelts. Sea otters eat sea urchins, and sea urchins graze away kelps. Kelps form the basis of much of the food web and provide structure where many other species live. Many species of invertebrates and small or juvenile fishes find shelter in kelp beds.

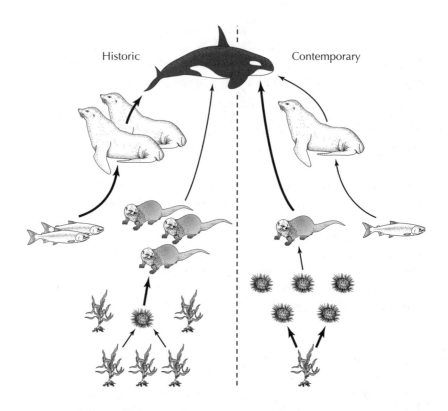

FIGURE 4.3 Marine food web along the northwestern coast of North America from Oregon to Alaska. The dotted line separates periods of original otter habitat (historic) and then otter removal and return (contemporary). The number and size of the pictures represent numbers of organisms (but not relative size of the organisms); direction of arrows indicates flow of food energy, from prey to predator; size of arrows indicates how much of each type of prey is eaten by the predator. At first, salmon and kelps (leafy underwater plants that provide protection for small fishes) were plentiful. After trappers removed otters, the number of sea urchins (small spiny invertebrate animals that eat kelps) grew explosively. After otters were protected, kelps returned, but then overfishing of salmon reduced seal and sea lion populations, causing killer whales (orcas) to switch to otters as prey.

With few otters, populations of urchins exploded locally, kelps were decimated, and the entire food web changed. As otters rebounded from overhunting in the past century, urchins decreased, kelps were replenished, and ecosystems were recovering. But, even more recently, otters began to decline again, and urchins are increasing. Otters are being eaten by orcas (killer whales). Researchers hypothesize that fishing by humans is reducing food supplies to seals and sea lions, decreasing their populations, and whaling has removed another food source of orcas. Orcas, no longer able to find seals and sea lions to eat, are switching to otters.[22] With fewer otters and more sea urchins, kelps are declining. Without kelps, the marine habitat is barren and biologically impoverished. This example illustrates the complex interactions in biological communities and the way human-induced alterations can cascade through food webs.

People harvest large fish from coral reefs throughout the world. Fish are removed for food or for the aquarium trade. Dynamite or cyanide is used to stun fish (and kill most fish that are not taken). Fish graze away algae that grow around corals. Once fish are removed, algae smother corals and corals die. Corals form structure and support for many species, so the loss of fish can cascade to fundamental changes in another marine ecosystem.

In both these examples of widely different ecosystems, extirpation of large fish has tremendous negative impacts on basic ecosystem structure. The Northwest coast and coral reefs are just two examples to illustrate the point; similar changes in ecosystems around the world have been caused by long-term effects of removal of large marine animals.

Regardless of utilitarian arguments for maintaining species, moral arguments are strongest. Few people think that causing a species to become extinct is acceptable. A jungle in Panama with no calling frogs or a forest in Guam where bird songs are silenced is wrong in some deep sense. Species extinction, leading to a biologically devastated world, is a tragic legacy we will leave for future generations.

Scientists estimate that there are over 10 million species on Earth,[23] and half of them may be extinct in the next 50 to 100 years if we protect only 10 percent of the native areas worldwide.[24] This projected number of species extinctions is an estimate because we have not even come close to cataloging all species. The way to estimate rates of extinction of less-known species is to account for the number of unique species per unit area in each habitat type, and then determine the rate of loss of that type of habitat. Most undescribed species on land are either insects or plants, and more species of insects are present than of all types of other animals combined. Two other habitats on Earth that contain tremendous numbers of unique species are coral reefs and the deep ocean. Pressures on coral reefs have already been discussed briefly; almost 60 percent of the world's coral reefs are threatened by human activities. Nutrient enrichment, fish removals, and global temperature increases all are causing great harm to coral reef biodiversity.[25] Far less is known about animals that inhabit the deep ocean and how human activities influence these animals.

More undescribed species occur in tropical rainforests than in any other single habitat on land. To estimate global biodiversity, estimating the number of undescribed species in tropical rainforests is essential. Biologists go to an area of rainforest and count all unique species of insects or plants. Then they find another area and count how many novel species occur in that area. This process is repeated for a number of areas. Once they determine the number of unique species expected in each general type of tropical forest as each new area is sampled, they can estimate the global number of species. The number of expected new species in each area (per square mile or square kilometer) is scaled up to total area of that habitat on Earth to estimate the total number of species on Earth. If this process is repeated for all poorly described, high-diversity habitats (particularly tropical forests, coral reefs, and deep oceans), and numbers are added to the already relatively well characterized numbers of species from temperate terrestrial habitats (still, new species of insects are being described everywhere), the total number of species on Earth can be estimated.

Extinction rates can be calculated in a similar fashion if we know the rate of destruction of individual habitats. The rate of habitat loss (area per year) is multiplied by the number of unique species per unit area to yield an estimate of the probable number of extinctions per year. Of course, these estimates are rough, but is it any less of a tragedy if 5 million rather than 6 million species are destroyed? The fact that so many species will be lost is what disturbs me.

The number of unique species in each unit area of tropical rainforest far exceeds that in most other habitats. Scientists know that if we decrease habitat area by 90 percent, at least 50 percent of the species will disappear. Current estimates are that 14,000 to 40,000 tropical forest species become extinct each year.[26] At current rates of deforestation,

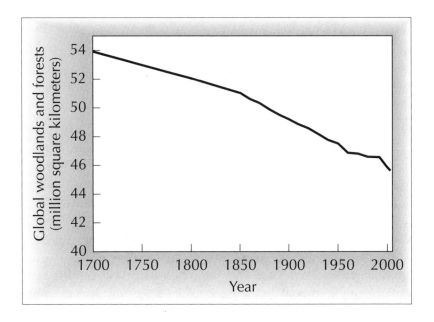

FIGURE 4.4 Global decrease in amount of forest and woodlands on Earth from 1700 to 2005. The decrease in forests has implications for greenhouse gases, biodiversity, and wood production.

humans will probably destroy at least 90 percent of tropical rainforests. Thus the majority of species on Earth will cease to exist in our lifetimes.

Mature forests are very likely to be cut down because wood is so valuable. For example, in the United States, only a small percentage of ancient forest in the Pacific Northwest remains. Despite these areas being home to endangered species, logging companies are working hard to get at the few remaining tracts of ancient forests. Similarly, tropical rainforests are under tremendous pressure for their hardwoods. If additional ancient rainforest is destroyed, an even greater proportion of species on Earth will be driven to extinction. Global loss of old-growth forests, about one-eighth in the past 200 years (figure 4.4), is partly related to tropical forest destruction. Deforestation in temperate zones has less of an effect on extinction, because there are fewer unique species per each unit area of temperate forest.

We know that it takes centuries for forests to recover, and millions of years for new species to evolve. A huge area of forest can be cut down in a year; associated species that go extinct will never return. The length of time for biodiversity to recover from deforestation is yet another example of environmental damage for short-term gain that takes much longer to heal than to cause.

Is Preservation Possible?

The current rates of extinction exceed the rates of evolution by at least 1000 times. Recovery of total numbers of species from previous mass extinction events documented in the fossil record took between 10 million and 100 million years.[27] The current species extinction crisis is directly related to habitat destruction by humans, which has greatly accelerated extinction rates. Our impact on Earth's species has been equaled only by ancient disasters preserved geologically, such as a giant asteroid slamming into Earth; humans are a catastrophe for most of their fellow species on the planet.

It is not possible to preserve just a few individuals of each species (the zoo and small preserve "solution") because the probability of extinction increases drastically with smaller populations. The extreme increase of extinction probability has been termed the "extinction vortex," and several processes feed back to cause extinction.[28] First, small populations are more likely to be wiped out by natural disasters and diseases. Second, fragmentation of populations can make each individual population much more susceptible to extinction. Third, smaller populations are subject to deleterious effects of inbreeding. Fourth and last, smaller populations are less able to evolve to accommodate environmental change (for example, altered climate from global warming). Humanity is both decreasing and fragmenting natural populations, while accelerating environmental change. Humans influence all four processes that feed into the extinction vortex, drastically increasing the probability that individual species will go extinct.

Species extinction cannot be reversed. Thus we have one more example of the incredible momentum of human effects on Earth's environment. It will take *millions* of years for new species to evolve to replace those that we have destroyed. This assumes that there will be habitat for them to inhabit (more native habitat than there is currently). Species extinction is stealing the natural world from our grandchildren. The cynic in me thinks that if our grandchildren care no more about the Earth that supports them than we as a global society now do, they will not know what they are missing. If our grandchildren do care, our callous disregard for species will be unforgivable.

Complex Interactions Between Ecosystem Health and Human Well-Being

The World Health Organization recently released a "Millennium Assessment" of the effects of ecosystem degradation on human health.[29] This report documents complex interactions between human health and the state of the world's ecosystems. Many changes noted in this volume are likely to ultimately encourage the spread of human disease and an increase in malnutrition. The global transmission rate of diseases has vastly increased because of rapid transport of humans, plants, and animals. Ecosystems are stressed, so native animals remaining in them are more susceptible to disease. All these ingredients increase the probability for catastrophic diseases to sweep the globe, causing massive human suffering.

An example of unforeseen negative health impacts is the emergence of the Nipah virus. Forest clearing in Indonesia drove bats carrying the disease to nearby Malaysia, where the virus infected pig farms and then humans. In Malaysia, people in contact with infected livestock contract the disease, and about one-half of the people with symptoms die. Similarly, AIDS is thought to have infected humans from other primates as more Africans moved into forests and came in closer contact with infected primate populations.

Livestock production in general has led to increased transmission of diseases from animals to humans. Recent concern over the spread of avian flu is but one example. New flu varieties repeatedly arise in areas where pigs and poultry live in close contact with people. This is why we need a new type of flu shot every year. With more people producing more livestock, living in closer quarters, and rapidly moving domestic meat around the world, the probability for pandemics of deadly new diseases increases.

As people become more protein deficient and move into more wild areas, reliance on bushmeat (wild animals killed for human consumption) increases. More wild meat leads to a greater possibility of cross-species disease transmission, as well as increasing pressures on rare and endangered species. Forest clearance and global climate change may alter transmission patterns for diseases like malaria, AIDS, and Lyme disease.

Ecosystems help provide clean water; water-associated infectious diseases currently claim 3.2 million people each year, and 1 billion people lack access to clean water. Damage to ecosystems decreases the ability of river and lake systems to "self-purify" from human wastes that enter them.

Much of the world's human population burns wood and charcoal for fuel. The constant demand for wood and charcoal exacerbates deforestation. Smoke from such burning can cause respiratory problems. Deforestation not only leads to extinction of other species but also has attendant health risks for humans.

These are but a few examples of potential interactions of ecosystem damage with human health. While previous sections considered global environmental problems mostly in isolation (for example, water scarcity, agricultural pollution, ozone depletion), such problems interact in complex and often unpredictable ways. These complex, unpredictable interactions are yet another reason to be concerned about the momentum of human global environmental impacts.

What Response Will Counter Global Environmental Threats?

In today's society, often little direct cost is incurred by entities that harm the environment. Repair of environmental damage, even if possible, usually costs much more than it would to prevent damage from occurring in the first place. The fact that corporations and governments are not held responsible for the costs of environmental damage they cause is an important factor behind global environmental problems. Holding those who cause damage economically responsible for associated environmental and social costs they create is called "internalization" of costs. Unless humans are held directly responsible for the environmental costs they create, damage will continue.

All global environmental effects have intensified over the past 50 years and will take many years to reverse, when humanity ever decides to change the way Earth is treated. The reason I say "when" is that in a finite world, we will eventually have to stop exploiting the global environment at the rates we currently do. Most of these impacts were documented to occur on a global scale for the first time in the latter half of the twentieth century. The twentieth century was also the first time a large number of governments expressed the will to protect the environment. Is the will to protect great enough to counter the tremendous momentum of global environmental threats?

As has already been established in this book, whether environmental damage occurs in small or large areas, the general pattern that emerges is that it takes a short time to cause damage and a long time to reverse it. Agents of environmental harm usually do not pay the costs of their actions. Thus a short period of individual gain at the expense of the environment leads to lasting negative effects that are paid for by society as a whole. It takes a short time to detonate a bomb but much longer to repair the damage it causes. If society does not hold entities that cause environmental damage responsible, current trends of environmental damage will not change.

A first step in establishing accountability for global environmental effects is reaching awareness that large-scale perturbations are usually negative:

> Suppose you were to open the back of your watch, close your eyes, and poke a pencil into the exposed works. The almost certain result would be damage to the watch. This result is not *absolutely* certain. There is some finite possibility that the watch was out of adjustment and that the random thrust of the pencil happened to make the precise change needed to improve it. However, this outcome is exceedingly improbable.[30]

The burden of proof that large-scale changes will not have negative impacts should rest on those who are causing them. It is only fair that the costs of mitigating the impacts of perturbations on environments should be borne by those who cause the impacts.

Such disturbances to Earth as the greenhouse effect and deforestation will take a century to reverse, even if we immediately stop introducing greenhouse gases into the atmosphere and allow cleared land to revert to forests. Other perturbations, such as contamination by organic chemicals and fertilizers and ozone depletion, will take at least decades to reverse. Unwanted species introductions may never be reversed, and extinction will certainly never be reversed. Thus our actions are catapulting us into the future with a momentum that we may not be able to control.

If our activities so damage Earth that its ability to support us is substantially diminished, we may have some warning, but we may not be able to reverse the effects in our lifetimes. Thus action is required now to deal with environmental problems on a global scale. Effective action necessitates an understanding of basic human behaviors related to the way we use environmental resources.

The only thing that will redeem mankind is cooperation.

—Bertrand Russell

5

Survival on a Finite Earth

The Ultimate Game, or Why Human Nature Destines Us to Use More Than Our Share

We will have to work together as a global society to control our impacts on the planet; individuals, businesses, and societies only looking out for themselves without considering the global implications of their actions have bought us to where we are now. Social scientists and biologists have conducted considerable research on cooperation. This research bears directly on how humans will achieve the global cooperation required to sustainably inhabit Earth. Keeping in mind the tendencies toward resource use discussed already, how might we overcome some negative aspects of human behavior with regard to environmental issues and excessive resource use?

Game theory offers a way to characterize behavioral aspects of resource use by groups of animals, including people. Game theory is based on simple contests that indicate how players respond to various conditions to maximize their payoff; it is useful for determining when people will be selfish and when they will be altruistic. One of the first games used by game theorists was the Prisoner's Dilemma. This game is played by two "prisoners" who are questioned separately. If both prisoners do not confess, they both get a reduced punishment. If one prisoner "rats out" the other, he goes free and the other receives the full punishment. If both prisoners confess, both are punished. Repeating this game and altering the penalties allows exploration of why and when people may choose to cooperate.[1] Other games in which people make choices related to the payoff are also used in game theory, and some that illustrate behavioral traits relevant to the use of resources are discussed here.

Protection of the global environment will certainly require understanding the human dimension and why people cooperate. Game theory offers a potentially useful avenue for understanding how small groups of people interact. Aggregate behavior on the global scale is the sum of how many small groups interact. The strength of game theory is that it can characterize general human strategies for decisions related to resource use. Results of some game theory contests have been verified across many human cultures, making it an attractive way to approach shared behaviors related to global environmental problems. Broad generalizations of behavior are important because environmental values differ widely across and within cultures.[2]

Exhaustion of a resource to the point of its destruction may be the inevitable outcome when normal people exploit an open access resource. Alternatively, various strategies that humans use to cooperate can prevent the fallout from selfish resource use. Resources are defined here as anything humans use from the environment (for example, water, energy, species, forests). For the sake of simplicity, I also consider Earth's ability to absorb pollutants as a resource (for example, the capacity of Earth's atmosphere to withstand CO_2 emissions without drastic changes in climate). I describe how economists can assign values to ecosystem goods and services (treat them as resources with a monetary value) in chapters 7 and 9. Application of game theory to resource use rates is a potential avenue for bridging social science and environmental science to address environmental problems.

The First Environmental Game: The "Tragedy of the Commons"

Garrett Hardin's "Tragedy of the Commons" is a simple application of game theory to environmental resource use.[3] Crystallization of this concept by Hardin was the defining accomplishment of a distinguished, although controversial, career. The article outlining the tragedy has been cited in over 4000 scholarly works and by many more popular writers. Hardin worried about the influence of uncontrolled human population growth and felt that morality was the only solution to problems arising from excessive population. He wrote: "Ruin is the destination toward which all men rush, each pursuing his own interest in a society that believes in the freedom of the commons." Hardin viewed himself as a bioethicist and was so adamant about his belief that humans had the right to choose their own manner of death that he and his wife, Jane, committed suicide shortly after their sixty-second wedding anniversary.

The tragedy of the commons occurs when a sufficient number of individuals can exploit a finite open-access resource base. The example Hardin used was a village communal pasture ("the commons") to which each individual has access; each person gains the most benefit by putting more animals on the pasture (feeding more cows). Thus overgrazing occurs, even though it will ultimately destroy the pasture for everyone, because each individual has the most to gain in the short term by grazing more cows. Any person who sees the pasture being overgrazed and does not increase his number of cows pays a penalty because his few remaining cows get less food anyway.

Overexploiting the commons is the story of essentially all open-access resources on a global scale (in spite of claims that technology obviates the limits of resource availability)[4] because most global resources are open access and the human population is large enough to make global resource bases functionally finite. In chapters 3 and 4, I document that over the past 200 years humans have exploited resources at rates that have global impacts and that rates of use are accelerating. The Earth and its resources have become a commons for nations racing to exploit them. We have experienced the collapse of every major unregulated fishery that has been exploited by humans,[5] global deforestation rates remain unabated, we continue to pollute the atmosphere with carbon dioxide and other chemicals, and numerous species have been driven to extinction.

The concept behind the tragedy of the commons has existed at least since the Middle Ages; even Aristotle said, "What is common to the greatest number has the least care bestowed upon it." Human nature as formed by evolution (acting for short-term competitive success) may offer no way to escape the tragedy of overexploitation of resource use on Earth. Humans, through either biological evolution or early cultural evolution, do have a way out of the tragedy of the commons, however—by control of those who take more than their fair share. Hardin specifically claimed that regulations, laws, and incentives were necessary to avert tragedy. Social scientists have identified some additional mechanisms of controlling overexploitation that will be covered in this chapter as well.

Maintaining a resource by regulations and laws could include privatizing the grazing commons or imposing grazing fees. Ranchers who own their own pastures have high incentives not to overgraze them and destroy their long-term livelihood. The privatization solution would probably not work for the global environment, although each person could have ownership of a small amount of Earth's resources, and a corporate or governmental structure could be created to regulate sustainable use of these resources (an idea explored further in chapter 9). Imposing governmental grazing fees could control the number of cattle, unless people sneak their cattle onto the commons, in which case laws and punishment would be required. A global fee structure for overuse of global resources could be erected only if most countries on Earth agreed to be subject to the authority of an oversight group.

The tragedy of the commons has been averted in certain historical contexts without the specific remedies thought necessary by Hardin.[6] In general, this has occurred only when the resource is viewed as communal property. For example, grazing commons were widespread and sustainable across much of postmedieval Europe. Sustainable use was ensured by small communities that limited the number of people who could use the commons, as well as the intensity with which users could graze livestock. In contrast to arguments by Hardin, such limitation was achieved without formal regulations, laws, or incentives. Rather, informal social mechanisms were agreed on, which led to sustainable grazing practices within small communities.[7] Each member of the small community clearly understood the limits of the resource base and then formally or informally enforced the controlled grazing rates. Still, I do not think that informal social mechanisms will ever exert control over global resource use rates by a world community, given the conditions under which such control is successful.

The commons has been the subject of considerable research by social scientists; a recent report published by the National Research Council details the progress toward understanding how tragedy is averted.[8] These studies cover relatively small common resource pools and document the conditions under which societies succeed or fail in avoiding the tragedy. The tragedy of the commons applies mainly in conditions where resource users cannot communicate effectively and have no way to develop trust in each other or the management regime.[9]

Unfortunately, these conditions are mostly approximated in global environmental problems related to shared resources. Individuals or small groups cause the problems and have little way to communicate and create trust nationally or internationally, and there is no environmental management regime that spans the globe for a common purpose. Social research to date does not appear to have found a way out of the global tragedy of the commons, short of global agreements to manage Earth for sustainability.

Majority cooperation is required for global environmental sustainability because whenever a significant number of people are involved, some will think they deserve a larger share and some will attempt to take it. They will take more in spite of what is socially acceptable, or they will decide that they can reap greater benefit from taking a larger share than they can gain from the incentives provided to conserve the resource. Potential ways to encourage cooperation include agreeing on regulations against excessive resource use, taxation to discourage resource use, incentives to encourage clean industry and lower pollution, and permits to limit the total amount of pollution. Taking a larger share than others may be accomplished by skirting regulations, finding loopholes, or outright illegal behavior. Taking a greater amount of resource occurs because some individuals perceive the benefits of obtaining the resources to outweigh the risks of being caught. In the regulated grazing commons, a herder may sneak more cows on at night, pay the authorities to ignore his extra cows, or influence the political process so laws allow him special treatment.

Ironically, sustainable postmedieval European pastures that served as an example of informal agreements to avoid regulation were ultimately taken from small communities of herders. The system eventually fell apart because of increased control by a few wealthy landowners and the advent of large-scale farming.[10] Political changes accompanying the industrial revolution favored those who used more than their individual share. Those with money and power seized control of grazing pastures that were historically a community resource. This is a story played out repeatedly with open-access resources. Small fisherman cannot compete with large factory fishing ships, and individuals are not as efficient at logging small tracts of forests as are large corporations. Over time, the unregulated resource is depleted in an unsustainable fashion. The "tragedy of the commons" seems an apt parable of the human condition as related to global resource use. It explains how normal people acting in their own best interest can deplete any resource if enough individuals have access to it. The tragedy applies only if most people on Earth act selfishly and take a greater share than can be sustained over time.

Are Humans Selfish Cheaters?

Are all humans selfish cheaters? The short answer is "no," but a large proportion of Earth's population uses resources at per capita rates that cannot be indefinitely sustained by our planet. Whether deliberate or not, using resources at the expense of others is selfish. While human behavior is highly variable, and there are exceptions to almost any generalization about behavior, generalizations are useful when exploring global trends because it is aggregate behavior that matters, not the exceptions.

Selfishness is the basic assumption behind game theory. Assuming that individuals act in their own best interest allows prediction of the optimal individual strategy for solving problems. Psychologists use the phrase "behaving with self-interest" and avoid the term "selfish." This aversion is probably because "selfish" implies a person who concentrates on his or her own advantage or well-being without regard for others. Nobody wants to be thought of as selfish.

Individuals in the United States and other affluent countries have resource use rates that far exceed the amount that Earth can provide to sustain individual lifestyles into the future, given calculations of ecological footprint. On average, each person in the affluent society is behaving "selfishly" and taking more than his or her share relative to other less-fortunate people alive today and to future generations. To claim otherwise is to sugarcoat the truth. My children never tell me I am being "self-interested" when I take too much dessert. I am being selfish when I use more than my share of Earth's resources.

In game theory, the strategy of opting out of cooperation is called "defection," but here I refer to a decision not to cooperate as "cheating." If future generations of humans have equal rights to Earth's resources, and people in lesser-developed countries do not deserve to be born into societies where the standard of living is substantially lower than in other countries, then behaving in a "self-interested" manner is cheating somebody else out of resources. Thus the selfish act of not cooperating—specifically, when taking more than an individual's share of a resource—is cheating, even if individuals are not aware that they are cheating. This definition is consistent with the legal system in many countries: ignorance is not a viable defense for breaking laws. Cheating is taking advantage, whether or not the selfish cheater is aware of what he or she is doing.

Game theory is useful because it can be used to investigate situations when individuals will cooperate, when they will cheat, and when they will take action to stop cheating. Understanding the basics of these behaviors is the first step in devising solutions to the conflict between human behavior and a finite Earth. In general, an individual will not cooperate when the potential net benefits of cheating outweigh those of cooperating. Research on altruism in humans and other animals reinforces this point. If we can understand how people calculate these benefits, we can create incentives, disincentives, punishments, or other ways to encourage cooperation (for example, research into more efficient ways to use resources) in maintaining Earth's sustainability.

Animals, including humans, behave selfishly because of their evolved propensity to survive and propagate. All nonhuman animal behaviors that seem altruistic (unselfish) can be explained as successful strategies of selfish individuals driven by their genetic program.[11] Richard Dawkins described this concept well in *The Selfish Gene*.[12] Dawkins is an influential intellectual figure because of his dogged insistence on genes as determinants of behavior, along with his ability to convey evolutionary theory in popular terms. Dawkins's strident atheism has stirred significant controversy. His intellectual achievements (he holds an endowed chair at Oxford University and has won numerous awards) and power of argument make his points on genes and evolution difficult to ignore, even if one does not agree with his views on religion. Dawkins clearly explained how evolution selects for strategies that benefit the survival of genes, leading to a ubiquitous strategy of "you scratch my back, I'll ride on yours." He describes numerous cases in the animal world that were thought to be behaviors for the good of the group but can be explained more logically as behaviors that increase the probability of transmission of genes.

The most logical way to approach the problem of global environmental damage is to assume that most people are selfish and will "cheat" if they perceive net benefit. Even if not all people behave this way, the ones who do are the ones who cause the most problems. Humans are unique in that, as a species, we are able to realize that our best interests can be served by avoiding destructive, short-term, selfish behavior. We are also unique in our degree of cultural, in addition to biological, evolution. So for now, I discuss how humans behave, assuming they are mostly selfish, and in chapter 6, I discuss the evolution of behavior and exceptions to this assumption. Given the assumption of selfishness, under what conditions is mutual aid expected and workable?

Why Do People Cooperate, and Is Altruism Common?

Human behavior can be more complex than simple evolutionary selfishness or altruism (cooperation without regard for individual benefit). Human behavioral choices are probably not always driven by selfish motives, and evolutionary and psychological altruism (altruism that arises from cultural evolution or moral consideration by individuals) can be thought of separately.[13] In my view, if we approach environmental problems by assuming that people behave altruistically (preserve the environment) when they think such behavior serves their own best interest, then the chances of solving those problems are maximized. Here, I make the assumption that most players operate selfishly and do not distinguish between evolutionary and psychological altruism. To experts, my terminology may seem somewhat fuzzy, but it does not affect my result. The end result of the game is the same, regardless of the mechanism of self-promotion in game play. The end result is not changed because psychological altruism may lead some proportion of players to play more altruistically than others, but the remaining selfish players behave as predicted (in the current case, take more than their share of environmental resources).

Perhaps one solution to global environmental problems lies in a large number of people behaving as extreme psychological altruists. This is Hardin's moral solution to the tragedy of the commons. I do not think a dramatic surge in the number of people behaving completely unselfishly is likely, and I can think of no mechanism that would bring it about. If psychological altruism were to mitigate human global environmental impacts, I would be happy to have the fundamental assumption of this analysis proved wrong.

Humans can act against the cultural and biological evolutionary tendencies toward self-promotion and preservation. Some people behave both completely selflessly (saints) or completely destructively (suicide bombers). But psychological studies do not support the idea that uncritical altruism (completely cooperative behavior with little regard for personal cost) is common. Cooperative behavior probably has cultural and genetic aspects. Steve Schwartz, a prominent psychologist from Boston University, theorized that human altruistic behavior is influenced by (1) personal obligation felt to take specific helping actions (the obligation to be "good"), (2) the occurrence of situations that activate an individual's norms and values (social conditions where a person is expected to be "good"), and (3) critical evaluation of whether altruistic behavior is truly warranted in any specific situation.[14] Research verifying that altruism depends on an individual's value system supports these ideas. Schwartz considered these findings supplementary to genetic explanations.[15] Further research has verified that human cooperative behavior can be explained on many levels, from genetic to cultural, consistent with the view of Schwartz. In these cases, altruism is still beneficial to the individual at some level.[16]

Game theory with selfish players is applicable to global resource use rates because if a sizable portion of "players" behave selfishly, resource consumption dynamics stay the same. Unconditional altruists may remove themselves from the game, but the remaining selfish players still have the ability to use resources in an unsustainable fashion. In the example of the tragedy of the commons, even if half the people placing cows on the pasture agree to cooperate and hold the number of cows down, the remaining half will still overexploit the pasture. Thus I use game theory here to characterize human approaches to resource management.

The next three sections of this chapter discuss game theory in depth. Investigating conditions under which people will choose to cooperate is an exciting and active area of behavioral scientific research. For those who find the level of detail too great, the take-home message of the next three sections is:

1. People behave selfishly.
2. Under some conditions, cooperation is beneficial to selfish individuals.
3. People of all cultures punish cheaters, often at considerable personal cost.

These seem to be fundamental and deeply rooted biological or cultural human behaviors.

Game Theory: "Let's Share Your Resources, and I'll Use Mine"

Since John von Neumann (mathematician, pioneer of the modern computer, expert in quantum mechanics, and through his work on atomic weapons as responsible as any individual for bringing Earth to the brink of nuclear destruction) discovered that game theory can describe why people make economic choices, games have been used frequently as a model for human behavior that occurs in real situations. Game theory has applications ranging from devising military and economic strategy to describing evolution of animal behavior, including human–environment interactions.[17] Economists have used game theory to predict how humans will make choices about when to cooperate, when to have children and how many, and how to use resources.[18] Game theory can be used to provide indications of how human behavior will be related to resource use rates in a finite world because it offers a powerful tool to predict the way selfish behaviors dictate the portion of any population that will take advantage of others. Game theory is based on the idea that people will cooperate to conserve resources only when it is in their own best interest, and it establishes under what conditions this cooperation will occur.

When Do Humans Cooperate to Conserve Resources?

Cooperation among most humans (also sometimes referred to as "reciprocal altruism") will be necessary to reach the point of sustainable use of the environment. Individual gain is most often the prime reason for cooperation among animals. Applying the assumption of selfishness to human behavior allows cost-benefit analysis to determine a balance between cheaters and cooperators (that is, when altruistic behavior will occur and at what level).

Consider a scenario in which cheaters always steal resources and cooperators always share effort in acquiring resources. Cooperators continue to work together as long as others do, but they refuse to assist known cheaters. Assume that the resource base is a fishery; cooperators establish how many fish can be caught and still protect an area from overharvesting. In this example, a group working together can harvest fish with the least amount of effort, but they limit their take to sustain the fishery. Cheaters take as many fish as they can in any way they can.

If more cheaters than cooperators are present, most individuals spend their time distrusting others, protecting their acquired resources from cheaters, and thus fishing less successfully. If a few cooperative strategists are present, they will tend to find other cooperators (because they will interact with cheaters only once), and the joint strategy should become more successful because they will spend less time fighting over resources. Some balance between cooperators and cheaters will exist, however, depending on the costs and benefits of each strategy.

The North Atlantic cod fishery provides one of many cases in which cheaters won before the cooperators could control resource use rates. While the fishery was unregulated, cod were harvested relentlessly. More-efficient equipment allowed fishermen to take an ever-greater number of smaller fish, but fishermen needed to combine forces to purchase bigger boats. When Iceland tried to control takes from its coastal fishery, English fishermen were willing to risk armed conflict to take fish. Ultimately, the cheaters prevailed because too many people harvested too many fish, and the commercial cod fishery, for the most part, has collapsed.[19]

Some additional examples of cheaters who avoid regulations, go against social norms, or are willing to incur penalties include poachers who hunt off season, polluters who dump in spite of regulations, people who water their lawns during times of restricted use, people who take welfare payments they do not need, and people who kill endangered species for personal profit. Game theory allows us to explore why some choose to cheat and others choose to cooperate; ultimately, it provides a feasible approach to guiding control over global resource use rates.

If cheating has a high cost relative to the benefit derived from co-operating (for example, cheaters are shot by cooperators), the mutual strategy will be more successful, and cooperators will dominate—but not completely, because a cheater strategy will be more and more successful, the higher the proportion of cooperators. Conversely, if the cost of cheating is low and rewards are high, a higher proportion of cheaters will become established.

The idea that cooperation is the preferred behavior when cost is low and payoff is high has been documented for a variety of organisms, from bacteria[20] to humans.[21] Mutualisms (cooperative species interactions in which both species benefit) are common in nature and include flower pollination, seed dispersal for fruits, corals (a plant and an animal), lichens (a plant and a fungus), legumes (plants with bacteria that help them get nitrogen), and bacteria that help animals digest food.

Mutualism allows some species to become very successful, and all known species have mutualistic relationships.

A specific example of the costs and benefits of mutualism is coral reef fish—some small fish clean parasites from large fish. Large fish go to cleaning stations and allow small fish to pick away their parasites. Cleaner fish are fearless and swim inside mouths and under gill covers of large fish to clean them. The small fish receive food, and the large fish are freed of harmful parasites. Both participants gain substantially from the interaction, with little cost. Of course, some species cheat; there are small fish that look and act like cleaners but actually bite chunks out of larger fish, and some large fish eat cleaner fish under certain conditions.

In flocks of birds and other groups of animals, warning calls may fall under the category of cooperation with small cost and large benefit. The alarm crier incurs little or no cost, but the benefit to the group is huge. As long as most individuals in the group are willing to make alarm calls, this behavior that benefits others can be very advantageous to each individual.

Many examples of cooperative behaviors in humans confirm that altruistic behavior occurs when there is little cost to provide help but immense benefit to the individual who receives help. Such behaviors include (1) helping endangered, hurt, or sick individuals; (2) sharing food or other resources; and (3) sharing implements or knowledge. Consider food sharing in a village of hunters; an individual who kills a large animal may share meat at very little cost because he and his family are unable to consume the entire kill before it rots. This same hunter may not be successful the next time but might receive meat from another cooperative hunter who is. Such examples illustrate how humans may assess costs and benefits and how this drives behavior.

A key aspect of game theory is that players respond proportionally relative to costs. Individual players calculate costs based on the proportion of other players with each strategy. Thus if the proportion of cheaters is high, the perceived benefits of cooperating increase. As a result, some proportion of humans will usually play cooperatively, and others will usually not, because the success of each strategy is defined relative to the proportion of each strategy in the total population. Proportional response is why punitive laws deter most, but not all, people from criminal activity.

These arguments hold even if a cheater sometimes cooperates, and vice versa (that is, strategy is context dependent). Each individual assesses costs and benefits of cooperating in each case, and a larger proportion of the population will cheat if the benefits of cheating increase relative to the costs. Most people would never rob a bank, but if a bag of money falls off of a bank truck and blows around, many people will keep the money they pick up, even though it is functionally the same as robbing the bank.

Individuals usually can be found who perceive the benefits of cheating to be great enough that they attempt to take more than their share. That a variety of strategies always occurs in populations is a function of natural genetic and cultural variability. Particular strategies may be successful, depending on the subset of other players interacted with, making chance a factor.[22] The proportion of cheaters overall, however, depends on how much is to be gained by cheating relative to the benefits of cooperating. In the global environment, cooperators strive to use resources at sustainable rates, and cheaters use more than their share. Although the existence of a proportion of cheaters seems to doom humanity to the fate of a few destroying the environment for the rest of us, there is reason for hope. Social scientists have demonstrated that humans have a deeply entrenched propensity to enforce fair behavior at considerable personal cost.

Reason for Hope (or at Least Revenge): The Desire to Punish Cheaters

Punishing behavior is so ingrained that individuals will punish cheaters at substantial cost to themselves. Punishment is common in all known human cultures. Even monkeys perceive social inequity as unfair, and nonhuman primates punish those they perceive as cheaters.[23] Neural scans of people suggest that we actually derive pleasure from punishing cheaters.[24] Experiments demonstrate that the right prefrontal cortex of the human brain controls punishment; interestingly, if the prefrontal cortex is disrupted, situations that are judged to be unfair are not responded to with punishment, suggesting the complex biological basis of punishment behavior.[25] Human punishment enforces cooperative behavior among unrelated individuals.[26] Several simple games illustrate that short-term cost-benefit analyses do not completely describe this human punishment behavior.[27]

Punishment forms the basis of most systems of law. The dissolution of social structure when punishment is suddenly not a consequence for asocial behavior is strong evidence for the partial deterrent value of punishment. Riots in cities, the streets of New Orleans after Hurricane Katrina, the conduct of some soldiers when they invade another country, and the simple fact that many people will cheat when they know they will not be caught—all these examples provide support for the threat of punishment as a deterrent to asocial behavior. Understanding the conditions under which people are willing to punish others, particularly at some cost to themselves, is a key piece of information in how people interact socially.

In the two-person Ultimatum Game, the first player is given $100 and may offer a portion to the second player. Players do not know each other and cannot exchange information other than an offer and a refusal. If the second player says yes, both players receive the proportion offered by the first player. If the second player refuses the offer, neither player gets any money. Since the game is played only once, the most rational response for the second player is to accept any amount, because even $1 is better than nothing, even though the other player gets $99. Refusing to accept a low offer is in effect punishing the first player for the low offer.

The surprising result of the Ultimatum Game is that most people offer somewhat less than one-half the money they receive, and more than one-half of the people who are offered less than 20 percent ($20) reject the offer. If people are truly objective about simply taking money in a game played only one time, they should take an offer of less than 20 percent. When the experiment is conducted with people of different cultures, ages, and socioeconomic classes, the exact results vary, but the gist is the same: people will punish those whom they perceive to be cheating them, even if it costs them to do so. In a study of 1762 adults from 15 populations across five continents, from college freshman to people living in a primitive society in New Guinea, mean offers ranged from 30 percent to 50 percent of the initial amount.[28] Apparently, players who make the initial decision on how to split the money understand that people have the tendency to punish selfishness, and they tend to make fair offers. More recent research on versions of the Ultimatum Game reveal how complexity alters outcomes. If multiple people can make offers, offers are smaller, perhaps because they perceive that the potential for future punishment is less.[29]

The Public Goods Game further explores the tendency of humans to cooperate and punish cheaters. In this game, four players each receive $20. Each player may donate to a common pool (the pot, in gambling parlance) but does not know what the other players are donating.

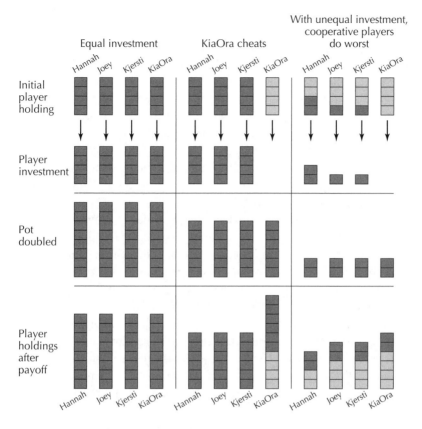

FIGURE 5.1 The Public Goods Game, in which all players start with the same amount and invest as much as they want. With equal investment (*left*), the pot is doubled and divided equally among players. A lone cheater does the best (*middle*). When all players invest unequally (*right*), nobody does as well as when they all cooperate. The lightly shaded chips are held back from the pot in the first round of play.

After each player donates or not, the pot is doubled by the experimenter, and then all the money in the pot is evenly divided among all the players. The best group strategy would be for all the players to invest the full amount every time, and then all would double their money every time. The best individual strategy would be to invest nothing but share the investments of others.

Three scenarios of the Public Goods Game can be illustrated by the first round of game play (figure 5.1). If all the players initially put $20 into the pot, they all end up with $40. If Hannah puts in $10, Joey and

Kjersti $5 each, and KiaOra nothing, then Hannah will end up with $20, Joey and Kjersti with $25 each, and KiaOra with $30. If Hannah, Joey, and Kjersti each put in $20, but KiaOra puts in nothing, after it is doubled and divided, Hannah, Joey, and Kjersti each have $30, but KiaOra ends up with $30 plus the original $20 that she held back. Thus, in the third scenario, KiaOra ends up with more money than if she had cooperated. When people really play the Public Goods Game, most initially invest about half their money; thus they play somewhat selfishly. If the game is played for 10 rounds, however, by the last round, most players invest little or nothing.

In a simple twist to the Public Goods Game, a player can impose fines on other players at a cost to themselves. So, for example, Joey can fine KiaOra $1 at a cost to him of 30 cents, and both fine and penalty amounts are removed from the game. If rules allowing fines are used, players fine those whom they perceive as cheating, even though nothing directly comes back to them. Players are willing to seek revenge on cheaters even at personal cost. When the game is played this way, all players tend to be more cooperative through the entire game, compared with when the game is played with no punishments. Results of studies based on this game reinforce the idea that human behavior includes some ingrained, long-term assessment of maintaining an even playing field for future contests.[30]

The Ultimatum and Public Goods Games illustrate that people behave selfishly in the short term, but they also calculate the long-term benefits of cooperation. In addition, these games demonstrate that when costs and benefits of cooperating are weighed, most players are somewhat cooperative. Other related games (a multiple-player version of the Prisoner's Dilemma discussed previously)[31] illustrate the same result as the Ultimatum and Public Goods Games.

Evolutionary and social science researchers continue to add complexity to games and to add factors that affect their outcome. For example, with many players, indirect cooperation also assumes importance: I will help you, if somebody else helps me. This type of reciprocity typifies the social structure of humans and leads to complex cooperative social systems.[32]

A central result of all games is that people are willing to punish cheaters at personal cost, and potential cheaters behave accordingly. Punishing behavior probably arose in small social groups of humans, where it was advantageous over a period of time to ensure that members of the group did not habitually cheat and use more than their fair share of the resources. Over much of human history, it has been easy to ascertain who is a cheater, and behaviors have evolved culturally or genetically to punish such cheaters.

One way that people signal and assess the potential for punishment by other members of the group is through the development of a social norm.[33] For example, people are more likely to throw trash into the ocean if they see others throwing trash into the ocean. This may explain why littering is less prevalent in the United States than it was 40 years ago; laws against littering have been enacted, but, more important, most people do not openly litter. Similarly, the willingness of Mexican citizens to conserve water is substantially greater if they perceive that others in the community are conserving water. Social aspects of water conservation in Mexico provide evidence that a social norm can control exploitation rates of a resource.[34]

The concept of development of infrastructure to facilitate implementation of social norms is called "social capital."[35] The formation of social capital happens in several ways. Relatively high social and economic well-being leads to social capital. Thus a social mechanism for enforcing group cooperation exists, but it depends in part on quality of life. In this sense, curing poverty may be important to global environmental protection. The concept of moral obligation to future generations encourages the formation of social capital.[36] Thus it may be too late to develop a social norm to protect the global environment if humans seriously damage the environmental capital that supports their quality of life, as well as any prospect of future generations obtaining a decent quality of life.

What Do All These Games Have to Do with the Real World?

The games discussed here are cases of "social traps," where players can be forced into decisions that may not be best in the long run. Four methods have been proposed to escape these traps: education, insurance, authority (for example, legal and religious systems), and converting the trap to a tradeoff.[37] These are interrelated escape strategies, and some will be difficult to implement at a global scale. Some of the fundamental properties of human behavior revealed by research on human decisions in artificial social traps can lead to insight on how to avoid the wrong decisions.

Willingness to punish others at personal cost is a basic feature of human behavior, at either the biological level (brain scans of chimpanzees illustrate this point) or the cultural level (punishment is a feature of most societies). Punishment may not be a completely effective way to alter behavior, as parents of any teenager can attest. Psychologists and economists have explored many ways to alter human behavior (for example, incentives, education, social pressure, taxes, user fees). Most of these routes to behavior modification entail a cost to society as a whole. Education requires financial support, taxes require individuals to pay, prisons require monetary support, incentives must be paid for. Behaviorists have demonstrated that incentives need to be applied continually to alter behavior, and the costs of behavior-altering incentives may exceed the benefits derived from behavioral alteration.[38]

Most methods of controlling behavior include costs to the "controllers." Exploration of when people are willing to punish others at a cost to themselves reveals the most fundamental calculations that humans make to determine if they are willing to cooperate to control behavior of members of the group who are perceived to be taking advantage of others. Estimated costs of controlling behaviors of others are part of the calculus of human society. When rates of cheating are viewed as unacceptable, the response of regulators is to increase the cost to cheaters if they are caught. Since most people weigh behavior on a cost-benefit basis, an increased cost of cheating should effectively decrease the proportion of cheaters. The proportion of jaywalkers would decrease substantially if the offense carried a death penalty.

Game trials used by researchers have described the behavior of only small groups of people. Small groups are how people interact with one another on a daily basis. Small groups of people interact within an ever-larger network of groups up to global society. Even if a portion of people do not behave selfishly, selfish behavior still will be the predominant mode of individual behavior. When summed to the entire globe, the basic results of game theory are consistent with the observation that humanity is exceeding the ecological footprint of long-term sustainability.

The global resource base is finite and can sustainably support only a set rate of use. Environmental impact is based on average rate of use; the greater the number of cheaters, the fewer the number of people who can be supported without damaging Earth's ecosystems. Unfortunately, human society has an escalating problem because increasingly more people exist, so the proportion of cheaters must be decreased simply to control the absolute number of cheaters. If the absolute number of cheaters increases, absolute environmental impact increases as well.

Stated another way, if total population doubles, there are twice as many cheaters even though the proportion of cheaters remains the same. For example, if one-quarter of the population cheats, but population doubles, the fraction of cheaters would need to be lowered to one-eighth of the population simply to keep the absolute number of cheaters the same. Thus penalties for cheating or incentives for cooperating would need to be made substantially greater. With a finite resource base, an increase in the number of people using the resource base will cause problems unless the proportion of cheaters can be decreased.

An additional aspect, equality, is important for controlling those who unfairly use resources. If sanctions are not applied in what is perceived as a fair fashion, rates of cooperation decrease drastically.[39] Thus laws that expose cheaters but do not punish them equitably can be worse than no laws at all. Incentives may also be ineffective if they are seen as unfairly rewarding a subset of the population.

Game theory predicts that some proportion of the population will decide that the payoff for not cooperating exceeds the possible costs for getting caught cheating. The more valuable the resource, the greater the penalties must be to stop cheaters. But as a resource becomes more and more limited, its value goes up accordingly, presenting an even greater incentive to cheat. Even with high costs imposed by regulations, the likelihood still exists that somebody will perceive the probability of being caught low enough to cheat.

People commit crimes even when there is a chance they will be killed if caught. Rhinoceros horns have a high monetary value that increases with rarity. Even though poachers know they may be killed if caught slaughtering a rhinoceros for its horn, poaching continues. If these poachers have starving families and the horn of a single animal can provide years of income, even the threat of death may not be enough to keep them from cheating.[40]

Governments behave in a similar fashion to individuals with regard to resource use and game theory. Thus we have nations cooperating to control the price and availability of oil and nations willing to go to war to ensure a continued supply of oil. As a resource, whaling has no significant benefits for most nations, so they readily agree not to exploit whales as a resource. Some nations are cheaters because the perceived value of whaling is greater than the costs of international censure.

Leaders of each nation strive to expand the amount of resources controlled and used by their nation. All behaviors described here for individuals apply to nations; governments cooperate when they think it benefits them, but some governments cheat. The more important a resource is to a number of nations, the less likely it is that global cooperation will occur. Penalties and punishments for cheating are much less defined at the international level than within countries because mechanisms for punishing cheaters are less established than in most small societal groups. Also, global social norms are less clearly defined than are those for smaller societal groups. Thus the ability to regulate global resource use rates is compromised.

Why Research and Education Are Important First Steps
Toward Solving Global Environmental Problems

Preventing people or nations from cheating and using more than their share of limiting resources requires general public acknowledgment of the limited nature of the resource base before sustainable resource use can be attained. That is, both the knowledge about the resource (research) and the transmission of that knowledge (education) must occur. Sustainable grazing commons in postmedieval Europe were managed successfully because community members were aware of (educated about) problems that would occur if people were allowed to graze more than their designated livestock or if more people were allowed to graze livestock. Research in this case presumably was the hard lesson the community had learned when the pasture was overgrazed at some time in the past.

In our modern world, people, governments, and corporations that use resources at greater than sustainable rates will never be controlled if most people do not perceive them as cheaters. Thus research and education are key components in efforts to conserve resources. Research is needed to convince the skeptical sector of the population that questions the urgency of protecting the global environment. Education from an early age and throughout adulthood is required to transmit information.

It was easy to detect cheaters in primitive human societies or even small communities in which resources needed for survival were clearly defined and accounted for. Accounting for global environmental damage and responsibility (research) and transmitting that information (education) is more complex in today's world. Let's consider three global environmental issues—the greenhouse effect, ozone depletion, and species extinction—in the context of costs and benefits for cheating, and consider the roles that might be played by research and education.

Game Theory, Research, and Education
Related to Three Global Problems

Earth's atmosphere and its ability to sustain life is a resource. One way the atmosphere sustains life is by regulating climate. Increased levels of greenhouse gases such as carbon dioxide from burning fossil fuels degrade the ability of the atmosphere to control temperatures at levels optimal for human inhabitation of the planet. The effects of more greenhouse gases are not necessarily immediately catastrophic for the entire human race, so the perceived costs of using the atmosphere to dump carbon dioxide are not extremely high.

The atmosphere is currently an open-access resource. Nations are unwilling to control their rate of burning fossil fuels, and the United States does not classify carbon dioxide as a pollutant. Most nations have been unable to agree to limit their rates of fossil fuel burning, even to current levels. Some nations completely refuse to enter negotiations; the United States, the largest single source of carbon dioxide pollution, is backing out of agreements that were previously made and claims that voluntary measures will be an effective method of control.

The atmosphere, being an open-access resource, is subject to the tragedy of the commons. As long as perceived benefits of overuse of the resource exceed costs, the resource will be used. Thus the United States continues to increase already massive rates of fossil fuel burning to sustain the lifestyle of its inhabitants, while claiming to be concerned about greenhouse gas emissions. China claims that raising its standard of living to that of developed nations is more important than any damage caused by the greenhouse effect, and emissions from China will grow, given its increasing population and rising standard of living (per capita resource use rates). Even if larger producers (for example, United States, Europe, and China) control carbon dioxide emissions, nations with ever-greater populations and increasing demand for energy (for example, India, Indonesia, and Latin American countries) are likely to refuse to control their emissions without substantial penalties for excessive carbon dioxide emissions.

103

Those unwilling to control greenhouse emissions seize on any uncertainty in the research on global warming. Until the overwhelming weight of evidence for global warming related to emission of greenhouse gases and the implications are understood by the public, governments will not be forced to act on the problem. Thus education and research are key components of regulation. Nothing in our prior cultural or genetic evolutionary history has prepared us to deal with global environmental problems. Research and education will be required to fill this void.

It was immediately clear that the destruction of the ozone layer could ultimately destroy much of life on Earth, in contrast to public uncertainty surrounding the environmental effects of greenhouse gas emissions. Potential damage from increased ultraviolet radiation is obvious to informed people in every nation of the world. Even though Donald P. Hodel, President Ronald Reagan's secretary of the Department of the Interior, once advocated sunscreen and hats to solve the problem, the public immediately realized the true gravity of the situation because little education was required to understand that a world where exposure to the sun for five minutes would cause a serious sunburn would be disastrous for all life on Earth. Once ethical considerations were included in addition to narrow economic analyses, the Reagan administration started pushing to control CFCs.[41]

A time lag of decades ensued between describing the problem and agreeing on a solution. Regulations to protect the ozone layer have finally been enacted, and emissions have slowed. Even though some countries or individuals may cheat and release ozone-destroying CFC compounds, the ozone layer may recover. In this case, research on ozone depletion and transmission of research results to the public might be sufficient to control the problem.

Species extinctions continue at an alarming rate because many countries do not view the costs of maintaining species (protecting habitats, reducing pollution, halting alien species invasions, and controlling species harvesting) as equal to the benefits gained from activities that ultimately lead to extinction (such as logging, slash-and-burn agriculture, urbanization, and other types of habitat destruction). Many nations have agreed in principal that humans should not cause species extinctions, yet extinctions continue at alarming rates in both developed and lesser-developed countries. Many perceive the benefits of activities leading to the extinction of species to outweigh the costs of protecting them. Here again is a gap between education and research.

We know the effects of many species extinctions and the reasons they are occurring. We know that species may be of economic value to society and that they cannot be replaced once they are extinct. The public has not been educated about the true dimensions of the crisis and what is required to preserve species.

In a general sense, when regulations to control resource use are developed, the costs of excessive resource use to society as a whole, as well as the potential benefits to cheaters, need to be assessed when setting the regulatory framework. Thus conservation biologists and other environmental scientists need to clearly identify the value of species and environmental attributes to be preserved, as well as the potential social cost for not doing so. Perhaps even more important, people with knowledge of the importance and uniqueness of particular resources must transmit that information to the public, to allow them to assess why it is in their best interest to comply with regulations and take steps to stop cheaters.

Environmental Education

Environmental education seems particularly essential, considering that a social norm makes compliance with regulations much more likely.[42] If most people agree with a regulation, a potential cheater will perceive a greater cost for cheating related to punishment from cooperators. Education needs to start at an early age and continue into adulthood. Effective environmental education should encompass a wide variety of subjects. Given the range of issues we face, one discipline alone cannot solve our problems, and education on environmental issues from a wide array of perspectives will prove more effective.

Information flow was relatively simple and comprehensive in primitive human societies or even small communities with easily accounted for resources. Educating adults about common-use resources in current times is complicated. Many adults use the news media as a primary source of information. Accurate information is required because accounting for global environmental damage and responsibility is complex. The news media needs to transmit accurate information and not create controversy over accepted scientific facts. This information is required to facilitate public awareness of the costs of not attending to environmental protection. Education also sheds light on who is benefiting from costs that society is paying.

In North America, Europe, and many other parts of the world, environmental education seems adequate at the elementary level. Most children understand the concepts behind species extinction, habitat destruction, toxic waste, and other environmental issues. Children know that causing extinction of a species or wanton destruction for financial gain is wrong. In secondary education, adequate environmental sciences (and other sciences as well) education begins to lag. Many adults do not have the scientific background to grasp the distinction between the greenhouse effect and global ozone depletion. Better secondary education and routes of education past that age are required to keep environmental awareness and concern in the public consciousness.

The environmental educator David Orr has proposed that a complete education allowing adults to make rational environmental decisions would include basic comprehension of (1) laws of thermodynamics; (2) basic principles of ecology, such as carrying capacity and energetics; (3) cost analyses, including complete accounting of costs (externalities) of use; (4) limits of technology; (5) appropriate scale (understanding how actions may have vast consequences); (6) sustainable agriculture and forestry; (7) steady-state economics (for example, growth is not always necessary or ultimately desirable); and (8) environmental ethics. He argues that our educational system has failed because educated people are causing tremendous global environmental damage. Education reform is necessary to allow complete education, not just create technicians. He notes that the architects of the Holocaust were products of what was regarded as one of the most-advanced education systems on Earth, and that incomplete education can lead to incomplete people.[43] Now the United States and Europe have some of the best-educated adult populations on Earth, yet they are disproportionately causing global environmental damage. With regard to environmental education, American and European adults are incomplete.

If adult education is derived in significant part from the news media, then the content of mainstream news sources reflects the source of facts that people pay attention to. Who decides what information is passed through the news? Fixation on trivial news stories is distraction. Advertisers also provide "information" on a regular basis. A mechanism for providing verified information on a regular basis that adults will pay attention to is a bigger challenge. It is interesting to me that a cornerstone of journalism is confirming the validity of sources, but there seems to be a general idea that established scientific facts are ideas that are open to dissenting viewpoints and that an untrained public figure should be given as much weight as the consensus of the entire community of science. Improving the flow of information between science and media will strengthen adult education. This book is another attempt to provide such education.

People who have been educated will understand who is cheating and using more than their fair share of resources, and who will be more likely to be willing to incur personal costs to help ensure equitable resource use by all. Earth's resources belong to all of humanity, present and future. Squandering this capital for the benefit of a few individuals, corporations, or nations does not make sense for the long-range survival of humans, and many other species. The longer we wait to control global explosion of resource use, the more difficult it will be to control. The momentum of global environmental impacts is not imparted just by the sheer large scale of global problems (such as the time it would take carbon dioxide to return to preindustrial levels). The momentum also has to do with behavioral processes that are deeply engrained in the interplay between biological and cultural evolution. Efforts to control resource exploitation are bound to fail if they do not consider human nature. We need to take advantage of existing behaviors, such as aversion to risk and punishment of cheaters. If we successfully control collective human behavior, we can win the ultimate game and our descendents will inherit a habitable Earth.

Why Humans Foul the Nest
Cultural and Genetic Roots Run Deep

People use far more than they need for survival and reproduction, lead-ing to tremendous rates of resource use in developed countries and a disproportionate burden on the global environment exacted by the wealthy. Because the root behavioral causes of resource overuse shape the relationship between humans and the global environment, consid-ering the roots of behavior is essential for the development of successful educational programs, as well as the establishment of policy to mini-mize global environmental impacts.

Edward O. Wilson of Harvard University is the father of sociobiol-ogy and one of the leading biologists of our time. He has established fundamental principles in the scientific study of ecology and has been a high-profile proponent of protecting the species on Earth. As a child, Wilson was fascinated by insects, and he claims that he "never grew out of his bug phase." It is likely that the study of complex interactions of social insects (ants and bees) led Wilson to think about the role of evolved behavior in human social interactions. When Wilson published *Sociobiology: The New Synthesis* in 1975, it generated immense interest and immediately provoked critical response.[1] Critics pointed out nu-merous cultures where some of Wilson's predictions did not hold. To-day, attacks on the sociobiological approach continue, going as far as to suggest that sociobiology is a "threat against religion, culture, and society."[2] Clearly, however, in spite of some exceptions, sociobiology is useful because it allows us to study the underpinnings of human behavior across cultures, and this approach is necessary to understand some of the deepest causes of our global environmental issues. Humanity is undeniably partially a product of biological and cultural evolution, and an evolutionary perspective provides a starting point for understanding the broad patterns in human behavior.

Application of game theory (chapter 5) is a sociobiological approach. One lesson from game theory is that there are conditions under which cooperation is a superior strategy, even for a selfish player. Cooperation can be explained purely by biological evolution, and I continue to explore the potential roots of other human behaviors as related to global environmental trends. My analysis, by necessity, is simplistic and does not cover some of the nuance and controversy associated with behavioral psychology and its relationship to environmental issues. For a detailed and thoughtful analysis of human behavior and the environment, I refer readers to previously published, more-extensive treatments.[3]

Why Worry About the Roots of Human Behavior?

People of all cultures consistently behave in ways that affect the environment, and consistency is partly why the best predictor of human behavior is past behavior. E. O. Wilson and Jared Diamond convincingly argued that human behaviors have both evolutionary and cultural roots and that social scientists and policy makers must understand these roots to solve humankind's problems.[4] Evolutionary and cultural bases of human behaviors are intimately intertwined with environmental impacts.[5] We live in a global society with a genetic program and cultural characteristics honed by hundreds of thousands of years of life in small groups. From the dawn of humanity until now, people and societies that most heavily exploited their environment for their own gain were often the most successful. Much of the momentum of global environmental problems is imparted by the catapult of our prior cultural and biological evolutionary history. Human behavior is deeply engrained, so it is hard to alter our aggregate behavior and this makes it difficult to change the present patterns of resource use and population growth.

The root causes of human impact on the environment have been attributed to economic systems that are out of sync with the realities of our modern world.[6] Some economists claim that evolutionary aspects have little to do with human economics;[7] yet others have viewed technology itself as evolving.[8] Optimists might argue that technology for minimizing environmental impacts could evolve more rapidly than the human impacts themselves. We can only hope this happens; it has not happened yet with respect to global environmental impacts. Understanding the implications of technological evolution still requires understanding the fundamental behavioral characteristics of humans. Thus I propose to take analysis of behaviors deeper and investigate how our nature and culture have forced us into a corner where our current economic systems are inappropriate for a finite world.

I present these arguments from an evolutionary point of view because it makes the most sense to me as a biologist, but the environmental consequences of human behaviors discussed in this chapter hold true regardless of how they came about. For those who prefer to attribute flaws and attributes of human nature to God, this scientific approach may still be useful. Discussion of how human behavior relates to global environmental issues is valid regardless of the origins of human behavior, but determining how hardwired some behaviors might be will assist the efforts to modify or control behaviors.

Managing the human-environment interface can be simplified if we aggregate the behaviors of humans that are most closely related to global effects into a few categories. I propose a list of fundamental behaviors that are deeply engrained by biological and cultural aspects. Many psychologists are not comfortable with classifying behaviors; such schemes are generally not used in modern psychology. As somebody once said, "Generalizations are usually untrue, even this one." As discussed in chapter 5, however, behaviorists are able to make general predictions about behavioral strategies based on categorization of two types of human behaviors: cheating and cooperating. Thus I categorize behaviors, while recognizing that other behaviors can come into play and that the "categories" of behavior listed here overlap.

Taking evolutionary psychology too far and categorizing behaviors into modules determined solely by genetics has been aptly criticized by David Buller, a psychologist and professor of philosophy.[9] Buller finds very few unambiguous studies of the evolutionary origins of human behavior. Data are weak because separating effects of cultural evolution from biological evolution is difficult: the two are intimately entwined. Approaching the issue of human influence on a global scale, however, requires finding a way to broadly categorize behaviors that lead the majority of people to have a negative influence on the global environment.

Defining types of behaviors may allow for selection of potentially appropriate approaches to modification of behaviors, prediction of which approaches will be successful, and potential quantification of environmental outcomes as a function of different management approaches. For example, construction of models of consequences of human systems, based on risk assessment, requires classification of behaviors.[10] This approach uses decision trees and classifications of the ways people react to perceived risk. While not all people act exactly according to the model, it provides a method to predict the general way that people will act. Risk assessment is a central tool in economic and resource management, and some of this assessment is based on classifying human behaviors. So let's explore some behavioral categories that could be related to global environmental impacts.

Fundamental Behaviors Related to the Environment

I propose that the most prominent traits influencing how we interact with our biosphere in ways that harm the global environment are the following:

1. A drive to reproduce
2. A need to control physical and biological surroundings to improve survival and comfort
3. A desire to display accumulation and consumption of resources
4. A tendency to cooperate with only closely related individuals

The tendency to cooperate with broadly unrelated individuals (for example, working together to solve environmental problems) and the propensity to control cheating and punish cheaters (as explored in chapter 5) are important to counteracting the four categories of negative behaviors. The four traits are listed in what I think is the probable order of decreasing genetic control. This order of listing is used because hardwired behaviors are the most difficult to alter. Changing the consequences of hardwired behaviors has a higher probability of success if methods are used that alter the effects of behaviors rather than the behaviors themselves. Behaviors more heavily motivated by cultural factors—such as consuming a specific resource for luxury purposes—are easier to modify because they can be more readily influenced by social pressure (such as laws, social norms, incentives, and peer pressure) and policies that influence costs (luxury taxes).

In some instances, amelioration of the consequences of behaviors is necessary. For example, controlling population growth by advocating birth control is more likely than by forbidding sexual intercourse or providing incentives not to copulate. Acceptance of birth control is a cultural characteristic, whereas the desire to mate is genetic. Although some countries have successfully decelerated birth rates (for example, China and Italy), and reasons for those successes may differ (strong governmental regulation in China, socioeconomic changes in Italy), I am not aware of any research demonstrating that frequency of copulation decreased substantially in these nations or others. Likewise, because the biological necessity to eat means we cannot survive without agriculture, we need to minimize the global impacts of that agriculture. No laws or education will keep people from eating. In other instances, actual behavior could be regulated. An example is the creation of laws and social conventions against using endangered species for nonessential purposes, such as the pet trade or adornment. So we need to understand how biological and cultural evolution led to specific behaviors and how hardwired these behaviors are.

Biological Evolution

Biological evolution occurs if characteristics related to increased survival can be inherited by offspring. Those characteristics are determined by genes, which can be passed to children. If organisms with a specific gene or group of genes have a greater survival rate and produce more offspring than those without, and if that ability is transmitted to their offspring (that is, it increases the reproductive success of offspring), the proportion of organisms with the gene or group of genes will increase in each successive generation. Differential success of characteristics that can be passed to offspring is called "natural selection." Natural selection leads to a greater number of individuals with the trait that conferred greater reproductive success.

"Variation" is the next essential ingredient for evolution. Genes must be reproduced very accurately so organisms retain all the complex functions required for them to survive. For example, the complexity needed to coordinate activities of the many organ systems in animals requires a high degree of fidelity in transmission of characteristics. Occasionally, there is variation in a gene, due to mutation or imperfect replication. Although this variation usually is neutral or harmful, sometimes it is beneficial. If the variation in the gene is beneficial, natural selection causes the proportion of the population carrying that gene to increase.

So how quickly does biological evolution occur? It depends on the species and number of individuals. Bacteria replicate very rapidly and have huge populations. If we expose a population of bacteria to an antibiotic, the probability that any single cell has the variation in its genes allowing it to resist the antibiotic is small. However, bacterial populations are approximately 1 million per 20 drops of water, and each cell can reproduce in a matter of hours. Strong selection to survive the antibiotic (bacteria that are not resistant to it die) almost ensures that each type of bacteria will be able to evolve resistance to antibiotics. This is why antibiotic-resistant bacteria are such a human health concern now, and efforts are under way to reduce the use of antibiotics to cases where they are truly needed. Evolution makes it virtually certain that disease bacteria will evolve tolerance to each new antibiotic if it is used often enough.

In contrast, human evolution is slower. Recent research shows that around 5 million years ago changes in only a few regulatory genes (genes that control activities of other genes) led to the evolution of humans from other primates. Of the 1056 genes analyzed for humans, macaques, orangutans, and chimpanzees, only about 60 percent have changed in the past 70 million years.[11] Some very important genes in animals (for example, those coding for reproductive behavior in salmon), however, have changed in fewer than 13 generations.[12]

One common error is to assume that biological evolution always leads to an optimal solution. This error seems particularly common in discussions on the biological basis of behavior. For example, when discussing altruism, people often describe a person who acted heroically to save another person even though it put her in mortal danger, but they rarely point out that if there is a crowd, they do not all jump into the freezing lake to save the drowning victim. Sometimes one of them does.

Why people behave the way they do is a complex subject. Some recent research suggests that humans have compartmentalized areas of the brain that have different roles in decision making. In this view, sometimes emotion trumps self-serving calculation. For example, love appears not to be a rational calculation of costs and benefits in many cases.[13] Natural selection demands variation, and higher animals have incredibly complex behavioral patterns, so it is no surprise that behavior does not always conform to a solution based on simple cost-benefit analyses. Nevertheless, the generalizations in this book are made because the empirical evidence points to the view that most of the time people behave in ways to maximize their individual benefit. It is average behavior across the globe that leads to our cumulative global environmental impact, not so much the exceptions. There is no way to change who we are biologically; we can only modify the expression of who we are via cultural means. Cultural evolution is thus the key to changing the way humans behave with respect to the global environment.

Cultural Evolution and Its Relationship to Biological Evolution

Cultural evolution is a fuzzier concept than biological evolution. Ideas are passed from person to person, and some become established for long periods of time (for example, some religions and city-states). The concept of cultural evolution led Richard Dawkins to propose "memes," or ideas that evolve in a cultural setting. Unlike biological evolution, ideas can change very quickly, can be spread from person to person, and have no strict inheritance pattern.

The mechanisms of biological and cultural evolution vary. Several important processes that influence change in cultural evolution include biases, nonrandom variation, and natural selection.[14] Humans are biased in what they observe and do not simply copy the behaviors of others, in contrast to biological reproduction where genes are copied almost perfectly. Whereas biological evolution is based on random variation, nonrandom variation can occur in cultural evolution. For example, a hierarchy of ideas can change all at once as variation in one idea cascades through others. Scientific progress is exemplified by this cascading nonrandom variation. The invention of new styles of music also occurs when one new idea influences implementation of existing forms of musical expression.

Rapid combination and change of ideas means that very complex constructs of ideas can be put together in new and previously unthought of ways. For this reason, cultural evolution is accelerating. The more useful ideas that are present, the more novel ways there are to combine those ideas. Around 3 million years ago (the date keeps being pushed back by new fossil finds), the first pre-humans appeared on Earth. Around 2 million years ago, our genus (*Homo*) arose, and use of tools began. By about 100,000 years ago, *Homo sapiens* began to spread across Europe and Asia. Only 10,000 years ago, art and agriculture appeared. Tools for agriculture took several million years to develop; fewer than 9000 years went by before writing and city-states arose. Over the past 1000 years, printing, industry, telephone, and radio have come about. In approximately the past 100 years (one-ninetieth as long as it took for art and agriculture to be invented), computers, plastics, air travel, space travel, television, the Internet, modern medicine, and numerous other innovations have occurred.

Information builds on information, accelerating cultural evolution. For example, an innovation such as controlling fire led to its use in preparing food, providing light, firing ceramics, melting metal, and so on.[15] Cultural evolution is far more rapid than biological evolution. The rate is accelerating with more people, more ideas, and more technology to move it forward. The tendency of cultural evolution to cause overshoot of the environment's capacity for support was acknowledged more than 25 years ago.[16]

Pre-humans were social species even before the genus *Homo* arose, allowing enough time for some biological evolution to occur. Biological evolution as a response to culture makes the entanglement between cultural and biological characteristics even more extreme. Humanity has reached a point where cultural evolution needs to be guided, and the consequences of biological evolution must be controlled. Control over cultural evolution will allow adaptation to life in a global society operating with genetic characteristics not suited for long-term survival. Up to this point in history, whole societies have paid the price for maladaptive cultural evolution, and other cultures have moved in to fill the void. Cultural evolution can lead to collapse of societies, and history has demonstrated that most civilizations have only a limited existence. Now, we live in a global society, and collapse of the global culture would probably be disastrous; there is no second chance if our global culture makes the entire Earth uninhabitable with environmental damage or global nuclear war. So how did humanity reach the point where cultural attributes could span the globe, and how do these attributes play off against our basic nature?

Humanity's sense of purpose has evolved first biologically and then culturally to ever-broader spheres.[17] A tradeoff exists between biological and cultural evolution (figure 6.1). Some aspects of human behavior, such as survival, have such a strong biological basis that biological evolution usually trumps cultural evolution. On occasion, however, cultural evolution does overcome the drive to survive; for example, although the practice is not widespread, *jumonji giri*, a Japanese form of ritual suicide, can be committed for cultural reasons. Cultural evolution has ever more influence on our behavior and survival. As we enter modern times, if the mitigation of global environmental problems is seen as essential to survival, cultural evolution could lead to the solution of global environmental problems.

The traits discussed here do not absolutely predetermine the behavior of every person at all times. Some individuals exhibit behaviors contrary to those listed here. Thus some people are highly altruistic and do much good in the world (or at least act contrary to their own self-interests). Other people believe so strongly that they are right—or are mentally deranged enough that they do not care about harming themselves—that they can cause tremendous harm to others, even though it costs them their own lives. Saints and suicide bombers, among others, act opposite to their selfish tendencies toward survival and reproduction (that is, biological evolution). Still, the global trends of resource use are driven by the aggregate behavior of all humans. Most humans are neither saints nor suicide bombers and tend to behave in a manner consistent with their own best interests.

Even if entire countries act contrary to their selfish tendencies, it may not be enough. For example, many European countries are making genuine efforts to reduce greenhouse emissions. The Global Footprint Network has calculated the ecological footprints of nations, just as they are calculated for individuals. This analysis suggests that in 2003 just six countries (United States, China, India, Russia, Japan, and Brazil) accounted for 48 percent of the total footprint of humanity (62 percent of the actual global capacity for sustainable inhabitation). All of Europe accounts for approximately 20 percent of the total ecological footprint.[18] Thus even if the continent of Europe, in aggregate, changes basic individual behavior or improves technology to substantially lower its global environmental impacts, it will have only a moderate influence on global trends. It is necessary to consider the most basic characteristics of human behavior applicable to the majority of people on Earth,

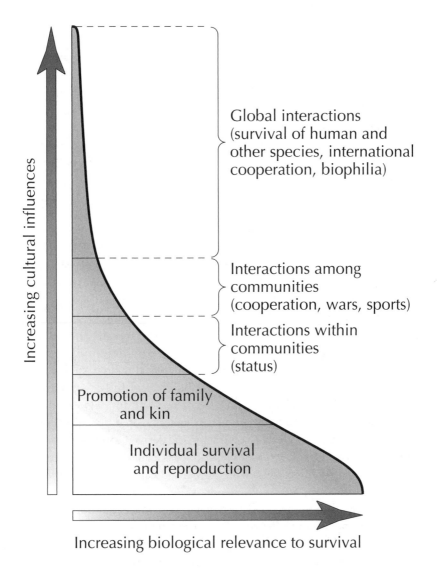

FIGURE 6.1 Tradeoff between biological and cultural evolution, indicating how cultural evolution will need to override biological evolution to solve global environmental problems. Some components, such as individual survival and reproduction, have such a strong biological component that cultural influences are unlikely to override them. Individual behaviors are discussed in more detail in the text.

and subsequently their societies, to explore what drives global environmental trends.

I turn to the four traits listed at the beginning of this chapter (reproduction, survival and comfort, excessive consumption, and cooperation with relatives) and one additional trait, biophilia, which has been proposed as a genetic predisposition to preserve our environment.[19] Finally, I speculate on how we can control effects of each trait.

Most People Want to Have Sex

The drive to reproduce, coupled with substantial food supplies and control of disease and predators, dictates that human populations will grow in the absence of reliable methods to prevent conception. This geometric population growth in the absence of regulation is a "natural" law. Natural selection strongly favors genes that encourage reproduction because such genes are more likely to be passed to future generations. Many species (for example, male black widow spiders and salmon) suffer death in a bid to reproduce successfully. Sex probably would not feel as good and be the goal of so many animals so much of the time if it was not a basic product of genes favored by natural selection. The sex drive is clearly a characteristic derived directly from evolutionary processes. The urge to reproduce is a characteristic shared by the vast majority of postpubescent humans, regardless of race or region of origin.

Most adult humans have the drive to reproduce, but not all reproductive events lead to offspring. Thus most population growth occurs in the lesser-developed countries, where access to contraception is limited, rights of women to make reproductive decisions are often not supported, and there may be financial and cultural incentives to have many children. Population growth has reached a plateau in developed countries where birth control is readily available and distinct financial incentives not to have many children exist. People in developed countries do not necessarily copulate less frequently; they just do not conceive children as often.

The urge to reproduce is such a fundamental biological behavior that controlling population growth by prohibiting sex is not a promising strategy. Increasing the standard of living and the status of women so that people want fewer children and providing the means to limit conception are more likely to be successful in controlling human population growth. The "catch-22" (lose-lose proposition) is that increases in standard of living correlated with lower fertility rates lead to greater resource use per person and tend to increase environmental impact. Thus controlling global environmental impacts solely by increasing standard of living to lower reproductive rate is not feasible unless the increased standard of living has substantially less impact than it does today.

Dominating Our Surroundings to Ensure Creature Comforts

Controlling our physical and biological surroundings to ensure survival and comfort is central to individual human success and domination of our planet. There is an immediate short-term advantage to controlling and using biological and physical resources, because humans who do so increase the probability that they will survive and reproduce. In this we are no different from all other organisms on the planet; evolutionary success demands attempts to convert environmental resources to promote survival and successful production of offspring.

Pleasure, comfort, and relief from labor (minimizing energy expenditures on needless activities) are outgrowths of selective pressure for survival. As a consequence, humans desire rich foods, moderate temperatures, and few physically taxing activities. Loading up on calories, avoiding extreme temperatures, and not working harder than necessary have historically led to greater survival and reproduction. Control of physical and biological surroundings has led to substantial increases in resource use to support our modern lifestyles, not to mention a wave of obesity in many developed countries. Anyone who has attempted dieting to control his or her weight can attest to the strong behavioral drive to consume more calories than are needed for survival.

People have displayed incredible ingenuity at controlling the physical and biological world, as evidenced by the continual technological advances in agriculture, construction, and medicine. This control is what has allowed us to be such a widespread and successful species. Our control of the physical world includes building structures to protect us from the elements, installing air conditioning and artificial lighting in those buildings, creating systems of transportation to move us quickly across the landscape, and developing the many modern conveniences that make life easier. All this control of the physical world requires the consumption of resources such as fossil fuels, trees, metals, and water. We all want to avoid hunger and other sources of discomfort.

Control of the biological world includes, among other things, evasion of and protection from predators, treatment of disease, domestication of animals and plants, and our current large-scale manipulations of all habitats on Earth. The clearest effect of this interaction is the exclusion of most other visible species, except those we have domesticated, from the areas in which we live. Thus we have modern cities with only a few urban macroorganisms (for example, rats, cockroaches, weeds, domestic plants, and pets) and numerous microorganisms. Most native species are gone. Even in the most primitive villages, people remove vegetation (so predators and enemies cannot approach undetected, for ease of movement, and maybe as fire protection), suggesting that control of other species is a hardwired human behavior with deep cultural and biological roots.

Many animals, including humans, remove parasites from their bodies, kill or drive away potentially competing animals, and encourage growth of organisms beneficial to them. Humans cook their food in part to decrease the propagation of parasitic diseases; they also cook their food to facilitate digestion, improve taste, as a social event, or as a symbol of power or wealth.[20] Historically and to the present day, cooking has led cultures to deforest large areas. For example, much of the Middle East was once forested. Several thousand years ago, humans used most of the trees for fuel and construction, and now the Middle East is a semidesert. This deforestation was driven by the human desire to obtain basic creature comforts.

To reduce resource consumption related to comfort and survival would require strict (thus probably unpopular) governmental control or more efficient means of obtaining creature comforts. Using technology to reduce impacts (changing the consequences of behavior rather than the behavior itself) is more likely to succeed than attempting to go against our evolved behaviors. Technological fixes include producing food more efficiently, employing best-management practices to reduce pollution associated with food production, using environmentally friendly construction materials, increasing energy efficiency, and many others.

Luxury Fever: Frenzied Affluenza Driven by Sexual Display and Hoarding

Desire to obtain and control (accumulate and consume) resources far exceeds the requirements for survival and comfort in many cultures. Overconsumption may have been part of what led to the collapse of the Mayan culture.[21] Aristotle wrote, "It is the nature of desire not to be satisfied, and most men live only for the gratification of it." Why do humans desire more than they need? Extreme excess of resource acquisition, beyond that required for survival and comfort, is particularly prevalent in the United States, but it exists in many other societies as well and has been termed "luxury fever" or "affluenza."[22]

The strange thing about this escalated consumption is that wealth, once basic needs are met, has little relationship to happiness (subjective well-being).[23] People think they will be happier if they are richer, but psychological studies have shown that they are not.[24] Data in the United States since the 1970s underscore the lack of correlation between absolute wealth and happiness. The Pew Institute has surveyed around 3000 people each year since 1972 with the question "Taken all together, how would you say things are these days; would you say that you are very happy, pretty happy, or not too happy?"[25] The reported level of happiness has changed little since the beginning of the study (figure 6.2).

Data in panel B of figure 6.2 suggest that happiness is related to wealth; people with higher incomes say they are happier than their lower-income counterparts. In absolute 2005 dollars, however, the lowest-income households in the United States have about as much now as the middle-income group had in 1972. Similarly, median household income now is about the same as the income of the wealthiest one-fifth of U.S. households in 1980. But happiness has not increased with absolute wealth. The explanation for this relationship that makes the most sense to me is that happiness is a function of how wealthy you are relative to other people. This finding might hint at the cause of desire for wealth beyond the provision of basic needs.

I doubt that excessive consumption is attributable to a single cause. Two major factors could be partly responsible for excessive consumption. First, I propose that, for some people, luxury consumption is behaviorally rooted in the need to attract mates. Second, it may be an outgrowth of the need to conserve resources during good times to survive in bad times (hoarding behavior). Both these causes may need to

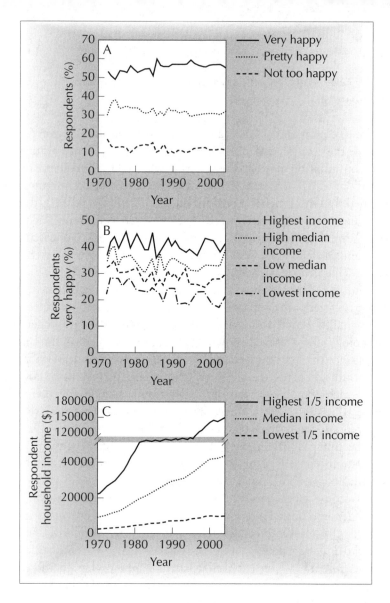

FIGURE 6.2 Level of happiness, as represented by the percentage of respondents reporting that they are very happy, pretty happy, or not too happy, in the United States from 1972 to 2005, as a function of (A) time and (B) relative income level. Household income in the same period (C) is adjusted to 2005 dollars; the break in the y-axis accounts for the very large increase, from $50,000 to $200,000, in the income of the highest one-fifth of income level. Note that by 2004, the lowest one-fifth of income level had almost reached the level of the median income that people had in 1972, but had not become any happier. Similarly, the richest people have become over seven times richer but remain at the same level of happiness. (Happiness survey data courtesy of the Pew Institute; household happiness from the U.S. Census Bureau)

be addressed to control global impacts. This may be fertile ground for environmental behavior research.

ATTRACTING MATES

In some species, direct physical contests are used to select mates; dramatic battles between males for mates are the core of many television nature shows. Combat for mates is not common in modern human societies (except, perhaps, ritually in the form of sports). Therefore, I propose that, in the modern world, control of resources is a characteristic that increases the probability of attracting a mate. Television and other media provide evidence that displays of wealth are enticing to the opposite sex. Many advertisers capitalize on the human desire to convey an impression of wealth by trying to convince potential buyers that their products will improve the ability to attract sexual partners. They create an artificial demand and "need" for products because owning them will supposedly improve image. The need to project self-image may ultimately drive behaviors that are not directly related to the attraction of mates;[26] logically, the behaviors could be rooted in the desire to attract mates. Acquiring an expensive, prestigious car may be directed toward impressing neighbors and friends, but the desire to project social status could arise originally from the need to attract mates. Culturally and probably genetically, the desire to obtain and consume resources is intensified in modern society.

One counterargument to the idea that the acquisition and use of resources enhances reproductive attractiveness is that people in countries with fewer resources have higher reproductive rates. This argument is incorrect because attracting a mate is based on relative, not absolute, comparisons of potential mates. The wealth of men is positively correlated with babies produced when analyzed within a country[27] (although social status does not always correlate with frequency of copulation, so other factors may come into play).[28] A positive correlation between fecundity and wealth across countries is not expected for other reasons; people in developed countries have fewer children due to greater availability of birth control, higher survival rates of offspring, and less need for children to work the land or support parents when they get older.

Influential and powerful people are apparently more successful at spreading their genes, so the desire for wealth and power could be a biologically beneficial natural trait. A genetic analysis of Y chromosomes in 2123 men in Asia from the Pacific to the Caspian Sea proves this point.[29] A certain Y chromosome occurs in approximately 80 percent of the men, or about 15 million people. The chromosome's pattern in variation indicates that it originated in Mongolia approximately 1000 years ago with Genghis Khan. Genghis Khan was one of the most powerful and influential men in human history. His power, evidently, resulted in tremendous reproductive success.

A relative comparison of the wealth of potential mates leads to a positive-feedback loop that continually increases consumption. Display of control over, and consumption of, resources by men and women escalates (contributing to luxury fever) because the excess appropriation of resources is a sexually selected characteristic. In a society in which standard of living is high, it is not enough simply to display control over sufficient resources to ensure the survival of you, your mate, and your offspring. You must control more resources than are controlled by your potential competitors for mates to make an attractive display. Thus a man or woman who is very successful and considered a good potential spouse in rural India could move to a large city in Australia, where he or she would seem poor by comparison and have more difficulty attracting a mate. With the globalization of Western television and movies glorifying wealthy lifestyles, the basis for comparison has escalated in all countries.

If luxury fever results from a need to attract mates, acquisition of resources is a result of sexual selection (natural selection of characteristics based solely on the ability to increase the opposite sex's desire for copulation) and an outgrowth of caring for offspring. Control and display of consumption of resources can be used to attract mates in species other than humans (for example, territorial behaviors to attract mates), and hoarding resources can give offspring an advantage.

Sexual selection is a well-defined biological trait of many animals. It can lead to secondary sexual characteristics that attract the opposite sex. Sexually selected characteristics and behaviors include colorful plumage on birds, mating displays and colorful markings on fishes, and defense of territory for many animals. Humans have their own unique sexual displays. Men have facial hair and large penises (for mammals our size), and women have relatively large breasts and gluteus maximus muscles. Members of the opposite sex prefer certain sexual characteristics and mate preferentially with those who have the best display; genes for displays dominate the population through the process of natural selection. Very muscular men are attractive to many women, even though there is little relationship between physical strength and survival in to-day's world. These characteristics do not have to be tied to any other re-quirements for survival, and they may even entail a considerable cost: peacock feathers are difficult for males to haul around, but peahens sure do go for them. Few species acquire resources far beyond their needs for survival, although some do: bowerbirds collect materials to construct elaborate bowers to attract mates. Some animals attempt to hold territories that are not required for sustenance, but only to prove their strength to prospective mates. Humans are not unique in acquir-ing resources beyond what they need, they just take the behavior to ex-tremes (as with many other behaviors). Sexual selection goes further in humans and is influenced by social status. Display of wealth and con-spicuous consumption was described at least a century ago and has since been linked to evolutionary theories of sexual selection.[30] Money and power increase reproductive success.[31] These are not new ideas but merit further discussion because they are linked to high rates of re-source use currently associated with individuals living in developed countries.

Many factors may make a mate attractive. Evolutionary psycholo-gists have started with a basic premise that men desire women with strong reproductive potential, whereas women desire men who have higher status.[32] Data do not paint a very clear picture on these issues. For example, humans did not necessarily evolve in societies where men controlled resources (matriarchy may have been common in many prehistoric societies). Thus, over evolutionary time, women may have desired men who could provide food and seemed likely to pass genes that increased survival (for example, strength, speed, health, in-telligence, and symmetrical features). Both women and men also as-sess whether their mate will assist in caring for offspring, and thus mating choices may take many forms, depending on culture and bio-logical preferences.

Passing wealth to your potentially reproductive children and even grandchildren can improve their chances of successful reproduction.[33] Thus acquiring and controlling wealth may be selected for well past the age of reproduction because offspring contain half their parents' genes, so helping them reproduce can be a successful evolutionary strategy. Helping offspring reproduce is particularly beneficial if you are old enough that your own reproductive potential is low. Therefore, many people continue to acquire substantial resources late into life, well past the age at which they can reproduce. Such behavior could be so ingrained in culture and genetics that it continues well after genes and ideas have been passed to the next generation.

Selection to stop behaviors that manifest themselves after the age of reproduction is weak unless the behaviors decrease the survival of offspring considerably. This lack of selection after reproduction may be why health problems, such as cancer and heart disease, exacerbated by lifestyle, afflict older people at such high rates; natural selection has a substantially weaker effect after reproduction. Despite weak selection for post-reproductive behaviors, people continue to copulate, eat, and display other evolved behaviors well after they are no longer fertile. For much of human evolutionary history, most individuals did not live long enough to enter their post-reproductive years. People past reproductive age are stuck with behavioral characteristics that evolved to benefit younger reproductive individuals. Thus many people pursue wealth, dress lavishly, buy extravagant houses, and consume resources well after their children have been raised and moved away.

HOARDING RESOURCES

Few studies that I am aware of examine the human propensity to hoard resources. One major difference between hoarding behavior and consumption as a display of status is that hoarding may not have a visible comparative feature. By definition, hoarded resources are hidden. It does seem as though excessive shopping could be a manifestation of hoarding behavior gone awry.

There are several caveats to my idea that luxury consumption is based on sexual selection or hoarding behavior. The first is that people can take any behavior to extremes for reasons far removed from their evolutionary basis. Taking behaviors to extremes seems to be a particularly strong characteristic of cultural evolution. The second is that institutions may be responsible for luxury consumption. The idea that social structure evolves and shapes patterns of consumption has been in the literature at least since the late nineteenth century, when Thorstein Veblen published his book *The Theory of the Leisure Class: An Economic Study of Institutions.*[34] Societies can promote excessive consumption of resources; the Mayas, the Egyptians, the inhabitants of Easter Island, and many other societies put massive amounts of resources into religious structures. This is an example of cultural evolution taken to

extremes that can be harmful to society as a whole. Most societies today are not using a proportion of resources equivalent to that required
for Egyptian or Mayan pyramids. Excessive consumption driven by demand of the entire society is a form of luxury consumption not covered
in the preceding arguments, but this may be immaterial to global environmental impacts.

In modern society, an average person's desire to own and consume
more than he or she needs for survival and comfort will ultimately drive
aggregate global resource use rates. People in societies with the highest
resource use rates, the richest societies, have disproportionate control
over resources and display of excess consumption of those resources.
These same people also cause a disproportionately large amount of damage to global environmental resources.

"Us Against Them," or Why Rabid Sports Fans Abound

Cooperation is selective, and an essential part of the behavior is that it
involves only a limited group (chapter 5). This selectivity is the aspect
of cooperation that makes it difficult for humans to cooperate on a
global scale. Selective, or directed, cooperative behavior can be ultimately destructive to our biosphere. The most fundamental biological
level of cooperation is driven by kin selection, and the biological component of cooperation constrains who is cooperated with in a directed
fashion. In kin selection, cooperation is most likely to occur among genetically related individuals.[35] The more genes two individuals have in
common (the more closely related), the greater the evolutionary payoff
for cooperation. Kin selection accounts for behaviors that may appear
to be altruistic. For example, a sibling shares one-quarter of his genes
with his siblings. If an individual is not likely to reproduce because she
is too young, then it may be advantageous (from the perspective of per-
petuating genes) to help care for a younger sibling. Once the individual
reaches the age of successful reproduction and is able to find a mate, it
becomes more beneficial to produce offspring (because parent and offspring share one-half of their genes) than to help raise a sibling. The
more related to other individuals you are, the greater the evolutionary
incentive to cooperate with them.

An extreme example of cooperative behavior driven by kin selection is found in social insects such as ants and bees; all individuals in
a colony are closely related (descendants of the queen) and thus

cooperate fully. A looser manifestation of kin selection occurs with behaviors such as sibling care in birds and mammals. Even closely related animals will not always cooperate unless the benefits are substantial.[36]

Natural selection can also explain why humans cooperate within groups. Competition between groups can lead to the extinction of some groups and the survival of others. Individuals who cooperate within the group are more likely to survive than those who do not. Analyses suggest that human characteristics such as sharing food outside the immediate family, monogamy, and other reproductive features of humans can lead to situations where genes for altruism can be favored in groups when there are other groups that do not contain the gene. Apparently, these conditions were present for much of human evolution.[37] Maintenance of altruistic genes in this situation would require a definite distinction between groups (that is, a clear maintenance of the distinction between "us" and "them").

Kin selection explains why cooperative behavior is constrained. Directed cooperation can be problematic because it allows extremely successful groups of humans to dominate Earth but, by its nature, precludes complete cooperation with those who are considered different. In very primitive hominids with small family groups, "different" meant not genetically related. In contemporary society, "different" has come to mean not part of the same political, racial, or religious group. The basic biological and cultural evolutionary roots of this behavioral mode encourage the distinction of "us against them."

Cooperation is directed because it initially evolved as a way to help genetically related individuals increase the survival of their own genes. Cooperation is most likely to occur when small groups of cooperative individuals compete with other groups. Cooperation gives one small group an advantage over groups that are less efficient within-group collaborators.

ORIGINS OF COOPERATION

Humans have evolved as the ultimate cooperative species, with a greater degree of specialization of individuals and more success than any other species. Other highly evolved cooperative species (for example, bees) have only a few major types of specialization. For most of human evolution, people lived in small, related groups. Cooperation among members made these groups more successful and directly enhanced the survival of genes and cultural practices. The formation of social systems

128

that reinforced such cooperation strengthened and extended it and led to the cultural evolution of collaborative behavior.

Cooperation can evolve at any of several levels of organization:

1. Cooperation can result from kin selection.
2. Two individuals (even those of different species) can exhibit direct reciprocity, defined as cooperation between any two unrelated individuals that repeatedly interact with each other.
3. Indirect reciprocity arises when individuals interact with members of a larger population, but not necessarily repeatedly with another individual. Indirect reciprocity is common in human interactions but less common in other organisms; it leads to the evolution of morality and social norms.
4. Network reciprocity occurs when individuals interact with a subset of others in the group, but all members of the group are ultimately connected with one another.
5. Group selection occurs when individuals interact only within groups, but the groups compete with one another.

All these cases have fairly simple, well-defined conditions under which cooperation will evolve through biological or cultural evolution, depending on the costs and benefits of cooperating.[38] In both network reciprocity and group selection, groups of cooperators can be successful, and the distinction between cooperators ("us") and cheaters ("them") is all important.

Cooperation extended from small family groups to tribes and ultimately to nations. In all instances, cooperation seems to require the concept of "us against them" to be successful. This biologically and culturally evolved trait presently takes the form of family devotion, prejudice against outsiders, company loyalty, community boosterism, cheering for sports teams, and nationalism. The cooperative nature of humans may be a basic characteristic, because children as young as 18 months try to help others, and even young chimpanzees attempt to help other chimps.[39] Although I argue here that cooperation within groups has a basis in cultural and biological evolution, others have suggested that such groups form as a natural outcome of individuals trying to maximize their success in economic and other competitive arenas, with little reference to basic human nature.[40] Regardless, it is still "us against them."

A few individuals transcend the "us against them" mentality and truly try to assist all humankind; those individuals are the exception rather than the rule. Most people behave on the basis of self-interest, but cultural characteristics (such as the idealism of doing good for all) rarely override natural selection. Because we are concerned with the aggregate behavior of humans on a global scale, the ability of humans to behave altruistically without regard to the "us against them" mentality is approached in future chapters.

RAMIFICATIONS FOR HUMAN SOCIETY

One aspect of cooperation implicit in the concept of "us against them" is lack of empathy for less-related individuals, which allows us to more readily harm the environments that other people rely on. Most people in developed countries are relatively unconcerned about environmental degradation in developing countries, even if it is pollution tied to industrial production of goods that they use directly. Lack of empathy for unrelated individuals is illustrated more graphically by the way we view human deaths. When a family member dies, humans experience tremendous grief. When a friend or community acquaintance dies, people also grieve. But most people in large cities pay little attention when strangers die, unless more than one person dies (for example, three or four at once), or they die in a way we can empathize with (bizarre deaths also receive attention because they are difficult to understand).

The September 11 tragedy in the United States deeply affected most people in the country because many innocent people died in an unusually violent attack. Violent conflicts in other countries recently have resulted in many more deaths at one time. For example, in 1994, the Rwandan Hutus killed approximately 800,000 of their Tutsi countrymen over a period of 100 days, yet the reaction to these deaths was not nearly as great in the United States as was the response to the September 11 deaths. Lack of empathy for "them" as opposed to "us" is a simple fact of human nature.

Life can be more complex than simply "us against them." Today's allies can be tomorrow's enemies. This plastic behavioral response occurs because humans assess benefits to cooperating in individual situations, and they collaborate when they think there are advantages. This flexibility means that there is a potential for mitigating human behaviors such that people could globally cooperate to sustain Earth's environment. Achieving global cooperation requires simultaneous empathy for humanity and recognition of the threat to Earth's ability to sustain us in the fashion we desire. If people realize that we are "us" rather than "them" and learn cooperation on a global scale, "us" would become humans who cooperate to sustain our Earth regardless of race, religion, or creed, and "them" would be the environmental cheaters.

Can Love of Nature Counter Negative Human Behaviors?

An innate love of nature, called "biophilia" by Edward O. Wilson, has been deemed necessary to protect the natural world.[41] This human behavior has been proposed as a primary reason to hope we can counter negative environmental effects of our actions to control the biological world. The basis of this idea is that love of nature could be a fundamental part of human character, and that any solution to our environmental problems lies in exploiting biophilia. Humans must appreciate and love nature if we are to save it.[42] Thinking about the extinction of species can be deeply disturbing.[43] Our human society, unfortunately, is far from using biophilia to our advantage, if it even exists.[44]

Many generations of people have lived in urban centers with almost no direct contact with nature. Ever more people live in urban areas, currently about one-half of the world's population (figure 6.3). These people reproduce and are content to live in cities; the closest they come to wild areas is watching a nature show on television, visiting a zoo, or taking a trip to a local urban park. For many people, for most of human history, nature has been dangerous at worst and uncomfortable at best. Preindustrial native cultures may worship nature, but many of their gods are not benevolent. Control of the biological world, including protection from disease and predators and the invention of agriculture, has been central to human success on Earth.

Nothing in our genetic makeup prevents us from surviving if we are distanced from nature. A relatively small proportion of the human population cannot be happy away from nature. A number of comparatively well-off people who live affluent lifestyles in urban settings are strongly supportive of environmental protection. Being a wealthy urban environmentalist is functionally an inconsistent approach to environmental protection, since resources consumed living in luxury in a city have tremendous negative impacts on the environment as whole.

The historical trend has been toward increasing the distance between humans and nature, with an ever-greater proportion of people having less and less contact with the environment that supports us. Since 1988, the number of visits per person to U.S. national parks has

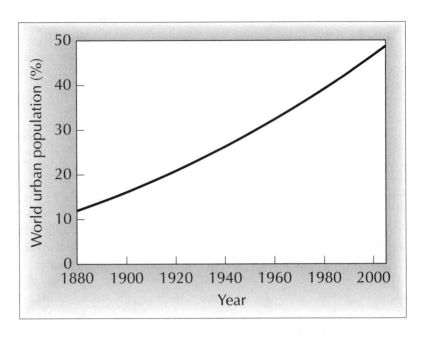

FIGURE 6.3 Expansion of the percentage of Earth's population living in urban areas from 1880 to 2005. Half of all people on Earth now live in cities.

declined by 20 percent. Coincidentally, the number of hours of television viewed, video games played, and DVDs and videos watched; the frequency of visits to theaters; as well as the rate of Internet use have increased. There was no change in the number of vacation days, and the effect was stronger with greater income. Increased oil prices also were related to declining park visits.[45] Apparently, the U.S. population is becoming more "videophilic" and transitioning from nature appreciation to sedentary activities involving electronic entertainment. This separation leads to the erroneous view that humanity does not need nature to support our continued existence and prosperity. Up to now, separation from the natural environment has not had an obvious negative impact on the success of the human species.

Products created by the entertainment industry may serve as one nonscientific indicator of the modern view of interactions with the natural world in the present and future. Most popular television programs and movies depict people with no relationship to nature. Nature shows are aired, but they attract a small proportion of the total television audience. Consider "starships" portrayed on television in which characters live for extended periods of time as they explore the galaxy;

audiences seem willing to believe that survival is easily sustained in a closed system for generations with no animals other than humans, some pets, and few plants.

Starships are still a fantasy; we are not able to construct an enclosed habitat that will support humans for a long period of time without external inputs. Inhabitants of the International Space Station need food and water that is carried up from Earth. If people think that a completely synthetic environment will be sustainable for long time periods in the future, they have no incentive to conserve the natural environment that sustains us in the present. Planning ahead for our own sustainable lifestyle seems a better bet for survival than relying on an innate genetic love of nature.

Potential Solutions to Effects of Ingrained Behaviors

I discussed four human behaviors in detail in this chapter: (1) the drive to reproduce, (2) a need to control physical and biological surroundings to improve survival and comfort, (3) the desire to display accumulation and consumption of resources, and (4) a propensity to cooperate selectively. Some solutions related to controlling the cumulative effects of our behaviors are straightforward, but others may be more difficult.

Mitigating the effects of the drive to reproduce is simple. Providing women with the right to make reproductive decisions and providing couples with birth control can reduce population growth rates. Effectiveness is enhanced if couples expect that their offspring are likely to survive. Contraception allows humans to avoid conception while expressing their genetically programmed need to copulate. Thus, given political will, control of the global human population is not difficult or extremely technologically challenging.

Survival and comfort are also ingrained, and most people will not act in a way to jeopardize their survival or even their comfort. Societies have voluntarily reduced consumption to free up resources to protect or expand their country—for example, during World War II. This type of cooperative sacrifice was driven by the "us against them" mentality, and death and destruction of a way of life were viewed as the penalty for not cooperating. Most humans do not view the global environmental situation as dire enough to require extensive individual sacrifices.

It is difficult to argue, without employing relativistic morality, that some people are more deserving of adequate nutrition and shelter than others. Thus environmentally benign technologies to provide sustenance and comfort to all people are necessary, and promotion of sustainability is also crucial. Education, incentives and disincentives, and cultural norms are critical to encourage people to use technological advances. Many existing technologies promote energy efficiency and less pollution, but they are currently not widely used.

Desires to accumulate and consume resources have strong social and cultural underpinnings, so establishment of social norms might be the best possibility for encouraging decreases in conspicuous consumption. If society views conspicuous consumption as an undesirable attribute for attracting mates, then the rates of conspicuous consumption will decrease. For example, although many wealthy women and men still wear furs from wild animals, the practice has become much less popular in developed countries because a significant segment of the population disapproves of this method of displaying wealth.

Status is in the mind of the beholder. In the Dr. Seuss story "The Sneetches," the Star-Belly Sneetches think they are better and the Plain-Belly Sneetches believe it.[46] First, the Plain-Bellies pay to have stars added to their bellies. Then, the Star-Bellies have their stars removed so they will feel superior. Eventually, the Sneetches spend all their money adding or removing stars from their bellies. Will humanity be like the Sneetches and spend all their environmental capital before realizing that conspicuous consumption is not a need or a fundamental right? We can only hope another method of sexual display is popularized before we use Earth beyond its capacity to provide support to the human population at a reasonable standard of living.

Many types of pollution are caused by more than one type of behavior. For example, the consumption of fossil fuels in the United States is related to maintaining sustenance and shelter, as well as to excessive consumption. The relative behavioral roots of these two types of consumption suggest that targeting excessive consumption will be a more successful method for controlling greenhouse gas emissions than will threatening the fundamental sustenance of the population. In the current political climate in the United States, excessive consumption has been elevated in status to a human right. The United States and some of its allies are willing to sacrifice youth in wars to preserve the flow of oil into our gas pumps. As long as people continue with excessive consumption as a social norm, the prospects for control of resource use rates are dim.

Global cooperation is necessary to solve global environmental problems. The only way to cooperate globally is to overcome the "us against them" mentality. If we view ourselves as a global society, then it will be possible to elicit cooperation of the majority of people on Earth to protect our biosphere. Religious, political, and popular leaders are needed who espouse global cooperation on environmental issues. Education is needed to show humanity the necessity of controlling our appetites. Given that who is "us" and who is "them" is a changeable characteristic, we should be able to broaden the circle and cooperate with all our fellow humans to sustain Earth.

Overcoming the tragedy of the commons requires cultural solutions because it is not likely to overcome our basic nature. An outline of conditions for sustainable use and avoidance of the tragedy of the commons is provided in a recent report from the U.S. National Academy of Sciences:[47]

1. Make the resource pool small, with well-defined boundaries and clearly delineated potential benefits to conserving the resource, and make the availability predictable.
2. Only allow small, well-defined groups with shared norms to oversee the resources. These groups need to be stable and interdependent, with informed leaders, similar interests in the resource, and low levels of poverty.
3. Ensure that the members of the group have sustained dependence on the resources, and require fair distribution at a slow, constant rate of the resources.
4. Ensure that institutional rules are simple and easy to understand, locally devised, and easily enforced, with graduated sanctions and fair administration.
5. Ensure that technology is available to exclude those from outside from using the resource, as well as the governmental structure set up so outside entities do not usurp the resource.

In general, all these solutions are based on cultural characteristics. How these cultural solutions to the tragedy of the commons will extend to controlling our global footprint is yet to be seen.

Give me where to stand, and I will move the Earth.

—Archimedes

Searching for Answers

Can We Achieve Sustainability, or Are We Screwed?

We cannot predict with absolute certainty what our future world will be like. Many possible scenarios exist, but it is impossible to continue to increase our total population and environmental impact indefinitely. Given the tremendous momentum of global environmental problems, humanity could be headed for some rough times if the trends documented in this book continue unabated. Earth's capacity to sustain us is finite. We know that our total population and resource use must be controlled; the question is how.

It will take decades, if not centuries or more, to mitigate the negative impacts of humans on the capacity of our Earth's environment to support us. Thus we could push our ecosystems past the point of being able to support us in comfort and not be able to return them to a sustainable state for many generations. As we consume Earth's resources, they become more rare and valuable; the more valuable a resource, the greater the incentive for a proportion of humans, including those who have power, to abuse that resource. The incentive also becomes greater for the rest of society to protect that resource, if they react in time. It is irresponsible to continue to behave as if nothing is wrong with the global environment. Nobody can say with absolute certainty either when we will use most of a resource or when we will irreparably damage it, but both are happening.

The people who most want to justify continued exploitation seize on any uncertainty. Equate scientific warnings about environmental perils to a mechanic telling you that your car's brakes may be damaged. Since repair may cost a lot, being a wise consumer, you visit other mechanics. None of the mechanics can tell you exactly when the brakes will fail, but all agree that failure is likely if the brakes are not repaired. The auto body repair-shop owner, ambulance-chasing lawyer, and mortician tell you not to worry because nobody can prove exactly when they will fail. You consider the fact that the auto body shop, the lawyer, and the mortician may not know much about brakes, and each has something to gain if you do not get the brakes fixed and have an accident. The rational response is to not argue with the mechanic over when the failure will occur, but get the brakes fixed. Why, then, do the media publicize those who disagree with hard-nosed, peer-reviewed, well-accepted scientific analyses of the state of our global environment? Why does a substantial portion of the public buy into this publicity?

A recent example is Bjørn Lomborg's book *The Skeptical Environmentalist*, which alleges that most major environmental concerns expressed by scientists are wrong or vastly overstated.[1] While major portions of the book were immediately debunked by respected scientists in the top scientific journals, it has sold a tremendous number of copies and is particularly favored by economists who have a philosophy that minimizes environmental concerns. Even though some of the public wants to be told that everything is all right, it may not be. Maybe even worse than ignoring the problem is purposefully not being educated about it in the first place. Perhaps arrogant ignorance is bliss, and those who do not believe environmental problems exist do not need to feel responsible. "Feel good today" is no prescription for tomorrow's problems. If you think there may be a problem with your car's brakes, it is not wise to avoid the mechanic to evade confirming the reality that the brakes need to be fixed.

The global environmental ozone-depletion problem is a good example of how scientific uncertainty can exist even in the face of certainty that a problem needs to be solved. Scientists cannot tell us exactly how large an ozone hole will result from a set amount of released refrigerants, exactly how long the effect will last, the exact health effects on ecosystems, or which people will die of cancer or develop cataracts.

Science cannot predict that any specific individual will get cancer from more UV exposure at a particular time. Still, the only rational action is to decrease the rates of release of ozone-depleting chemicals. But industrial polluters seized on the uncertainty and claimed that they did not have to address the problem if future effects could not be predicted exactly. Fortunately, the global public and the political system did not think the uncertainty warranted ignoring the problem.

Those who gain from polluting the environment with industrial wastes, automobile emissions, mining runoff, toxic chemicals, nuclear wastes, and many other pollutants often try to debunk scientific facts. These polluters usually hire scientists to publicly dispute the level of scientific certainty. The media love controversy and seize on these disagreements, often under a guise of unbiased reporting, giving the public a sense of doubt and lack of urgency about solving what clearly are gigantic problems.

We are building momentum toward what could be a catastrophic breakdown of global sustenance. No one can say if we are simply increasing the tension on the counterweight that powers the environmental catapult, are nearing the breaking point of the release mechanism, or have already released the catapult sending Earth into an unknown and unalterable trajectory toward a future crash. Such uncertainty does not mean we should keep ratcheting up the tension of the catapult because we do not know exactly when we will pass a point of no return. Continuing without seriously investigating global environmental consequences would be an optimistic, selfish, shortsighted, and foolish course of behavior.

It will be difficult for humans to sustain the ability of our planet to provide us a comfortable home. For sustainability to occur, we must reduce the rates of resource use, the environmental impacts of resource use, and the size of the human population; at the very least, we must hold all of them to present-day levels. Behavioral constraints (chapters 5 and 6) channel aggregate behavior to such a degree that it will be extremely difficult to use our resources wisely as a global society. Strong measures are required to reverse the path we are taking. Pessimism may be counterproductive, but realism is mandatory if we are to confront human nature and problems associated with sustaining a comfortable existence on our globe. In this chapter, I explore what it will take to use Earth in a sustainable manner, the potential avenues for attaining this goal, and what must be done to protect the global environment before the momentum of our actions sends us careening into a future nightmare.

Why Should We Protect Earth?

At various points in this book, I have made different arguments for protecting Earth. These can be summarized as (1) morality (it is the right thing to do), (2) sustainability (a subset of morality, it is not right to deprive future generations), (3) the precautionary principle (avoiding unpredicted catastrophe), and (4) economic benefit.

I am not an expert in philosophy, so morality is a difficult subject for me to cover, but I do believe that protecting Earth for its current and future inhabitants is the right thing to do. The precautionary principle is problematic because too much worry about the future could paralyze humanity, thus preventing environmental protection. Economists have argued that calculating costs and benefits is more practical for environmental decision making than using the precautionary principle blindly.[2] But costs and benefits can be difficult to assign in many cases where outcomes are highly unpredictable. Moreover, economic values are not the only important values to humanity. On the one hand, cost-benefit analyses have no bearing on moral decisions: if a national park will never generate as much revenue as mining the land would, is it wrong to preserve the park in perpetuity? On the other hand, cost-benefit analyses are useful when done correctly because they point out many cases where protection of the environment is economically desirable. I propose that when costs and benefits are difficult to assign due to lack of knowledge, more weight be given to the precautionary principle.

A discrepancy between cost-benefit analyses and the precautionary principle is illustrated by the way the U.S. government regulates drugs and food. In the case of drugs, the precautionary principle takes primacy: drug companies must prove that their product is effective and does not cause harm before it is released on the market. Herbal supplements, in contrast, are regulated as food. Only after credible evidence reveals that a herbal supplement causes harm does the responsible company have to stop selling it. The regulation of chemicals released into the environment is also based on cost-benefit analysis and usually is not governed by the precautionary principle. Part of the reason we have so many cases of toxic compounds causing widespread environmental damage is that companies are not required to prove, before they are released into our environment, that chemicals will cause minimal harm. Numerous companies understand the potential for unintended harm and financial responsibility and test their chemicals at their own expense before releasing them into the environment. Unfortunately, many other companies do not.

What Is Required to Take Care of Earth?

Julian D. Marshall and Michael W. Toffel argue that a hierarchical definition of sustainability should be based first on human survival.[3] They suggest human health as the next yardstick of sustainability, followed by lack of species extinction and recognition of human rights. Maintaining quality of life, values, beliefs, or aesthetic preferences makes up the last aspect of sustainability as defined by others, but they argue that sustainability should exclude this category. In general, I agree with their definition, but it must be recognized that few people will support measures that reduce their quality of life. Most people would agree that human survival, at the very least, should be sustained.

A commonsense set of recommendations for maintaining sustainability comes from Herman E. Daly and John B. Cobb Jr.:[4]

1. If it ain't broke, don't fix it.
2. If you must tinker, save all the pieces.
3. If you don't know where you are going, slow down.

With respect to the global environment, these guidelines mean that we should conserve the ecosystems we already have and not change them without understanding the consequences of doing so. The escalation of greenhouse gases caused by humans, and potential effects of these increases, are just one area where we are violating Daly and Cobb's practical recommendations.

Barry Commoner, an environmental activist and one-time presidential candidate, proposed several principles to be applied when approaching environmental issues that I paraphrase as:[5]

1. No action is without side effects.
2. Nothing ever goes away.
3. You can't get something for nothing.

These three "laws" apply to both global and local environmental issues. The first principle is a feature of the web of interdependence of all organisms and the reciprocal interaction of organisms with their environment. Because all plants, animals, and microbes are linked by the interactions among them (some not directly linked, but linked nonetheless), it is impossible to make a change that influences only one species. The second principle is actually a physical law based on the conservation of matter. The third principle is based in part on the first and second principles and in part on the laws of thermodynamics. You can't get something for nothing; no process is 100 percent efficient, so there is no perpetual-motion machine. These laws need to be kept in mind when finding solutions for ecological problems.

Given global impacts of humans and root causes of those impacts, I suggest the following specific goals to allow the sustainable inhabitation of Earth while maintaining some of the biodiversity and ecosystem functions that support humans:

1. Control the human population to at or below current levels.
2. In developed countries, decrease rates of resource use per person by one-quarter; everywhere, decrease the environmental impact per unit of resources used by one-quarter (for example, improve energy and water use efficiency, move to cleaner energy sources, increase material reuse rates).
3. Decrease rates of human release of greenhouse gases to levels that would allow Earth's climate to stabilize.
4. Halt or minimize release of CFCs and other chemicals that destroy ozone.
5. Protect our water from pollution by fertilizers, antibiotics, pesticides, and other toxins by using best-management practices.
6. Protect substantial portions (at least 10 percent) of all major habitats on Earth (for example, tropical rainforests, coral reefs, Arctic tundra, prairies) and selected "biodiversity hot spots" to support biodiversity and functions of these ecosystems.
7. Control the rate of human-caused species invasions.
8. Halt the massive rates of human-caused species extinctions.

Population control alone will not solve our problems because the relationship between rates of resource use and number of people is complex and difficult to predict.[6] A major step in reaching the goals I have listed is to use resources on Earth at a sustainable rate. Finding an exact definition of a "sustainable" level of resource use is difficult, but I will present a simple calculation of the *minimum* I think it will take to achieve this level.

For purposes of this calculated example, assume that the impact of resource use is accurately reflected by current energy use rates; this seems reasonable because economic activity, energy use, and environmental impacts are clearly interrelated (as demonstrated in chapters 2 to 4). Also assume that technological advances will make it possible to maintain current rates of total global resource use sustainably (that is, technology, such as renewable energy sources and increased efficiency, can reduce the environmental impact such that current global energy use rates become sustainable). Assume that the current standard of living in the United States and western Europe can be maintained with an energy use of one-quarter of the rates around 2000 if appropriate technologies are used (considered realistic by some)[7] that minimize CO_2 emissions.[8] Finally, assume that all people on Earth deserve a standard of living similar to that in the United States, Canada, and western Europe and that reaching one-quarter of the current consumption rates, coupled with more efficient use and less impact per unit use in these developed areas, accurately reflects the attainment of this standard of living.

With these very optimistic assumptions, 6.7 billion people could be provided with energy if the total energy used is the same as what was being used at the turn of the century. According to the U.S. Department of Energy, total global energy use is 381×10^{15} BTU per year (a BTU [British thermal unit] is an energy unit expressed in heat; 10^{15} is 1 thousand million million), and the total global population is 6.5 billion people.[9] The United States, Canada, and western Europe use 181×10^{15} BTU per year, with a total population of 778 million people. Now, 13 percent of the world's population uses 48 percent of the energy! This rate of energy use translates to 233 million BTU per person per year in these developed areas. Assuming that one-quarter of the current energy use in developed countries will provide an adequate lifestyle, one-quarter of the current rate, or 58 million BTU, are needed per person per year.

I calculated how many people can be supported by dividing total BTU used on Earth in 1999 (energy available) by BTU per person per year needed to sustain a desirable lifestyle. The total number of people supported at this energy use rate would be 6.7 billion people. Attaining this "minimum" level required for sustainability will be very difficult. To start, Earth's population will probably exceed 6.7 billion people near the time this book is published. If Earth's population reaches 10 billion (which seems likely), resource use rates will need to be cut by an additional one-third per person (17 percent of current use rates in developed countries). Other calculations using less-optimistic assumptions suggest that only about 2 billion people can be supported sustainably.[10]

Changes to an economic system to accommodate a fourfold increase in efficiency (one-quarter the energy use per person in developed countries) will be painful to companies and industries that currently profit from providing energy. Companies that benefit from status quo resource use patterns would battle tooth and nail against any attempts to make changes that would greatly improve energy efficiency and reduce their profits. In reality, increasing efficiency can generate economic activity by supporting industries that produce technology to improve energy efficiency (although economic activity from the supporting industries may or may not offset losses to industries that provide energy). For example, regulations requiring automobile makers to produce more fuel-efficient cars will lower oil company profits if less oil is consumed, but companies that produce parts for electric cars or hybrids may benefit.

If only a twofold increase in efficiency is attained (which actually sounds optimistic to me) and each person on Earth uses one-half as much energy as people in developed countries use currently, then only one-half of the current Earth's population could be sustained at current energy use rates. I repeat that my calculation is optimistic because it is based on the assumption that environmentally benign technologies will be adopted, allowing the current rate of energy use on Earth to be sustainable. Some obstacles to the considerable technological, social, and political changes must be overcome for these conditions to hold.

Let's consider sustainable water use rates. In the United States, people use about 95,349 cubic feet ($2700 \, m^3$) of water per person per year. If the standard of living of the entire world rose to that of the United States, and each person had the same per capita water use rates, we would use all the water that is geographically and temporally accessible on Earth (for example, some rivers flow where water is not useful, and runoff caused by flooding often cannot be used by humans). Now consider that Earth's population is growing. Increased demand for water means that substantial improvements must be made in the efficiency with which we use it. This potential scarcity of clean water will be further constrained by our ability to feed an increasing population. Higher levels of food production will require more cropland in areas that need irrigation. Increasing production of cropland will require more use of chemical fertilizers and pesticides that stress our aquatic resources to a greater degree.

These calculations demonstrate some potential difficulties in reaching sustainable resource use rates. They are crude, but both energy and water use rates illustrate that how we use resources, as well as how many people use resources, are important dimensions of global environmental impacts. How can humanity attain a sustainable future?

Linking Human Behavior to Reaching Sustainability

Altering human behavior is essential to mitigating our global environmental problems. Influencing behavior must be done with the pervasive aspects of human behaviors in mind. To review, these are (1) a drive to reproduce, (2) an ability to control physical and biological surroundings to ensure survival and comfort, (3) the excessive accumulation and consumption of resources, (4) a tendency to cheat (use more than a fair share of resources), and (5) the display of cooperative behavior only within a group.

The drive to reproduce is an integral part of human nature. The most successful population control is associated with access to contraception, improved rights of women to make reproductive decisions, and higher standards of living. Technology is available, and people with money can purchase contraceptives for those who cannot afford them. Unfortunately, providing access to contraception does not mean that it will be used. Many social factors may lead people to shun the use of contraceptives. Women's rights have expanded in many countries in the world. I believe that social structure rather than genetic factors causes men to deny women's rights. Social structure is difficult but not impossible to modify, however.

A greater standard of living is also occurring throughout the world, although many improvements remain to be made. Increases in living standard are accompanied by greater resource use rates. Without modifying resource use rates per person, there is no net gain in reducing environmental impact if population increases. Thus a transition from a low standard of living and high population growth rate to a low population growth rate and high standard of living may ultimately lead to greater resource use rates. Even though population control alone will not solve global environmental problems, it is an essential part of the solution.

The inability of humans to control their impact on Earth is highlighted by their inability to control human population. The problem of overpopulation has been understood by much of the public for at least 30 years. Loud-and-clear warnings were made, and considerable political interest has been generated since at least 1968.[11] Cheap and reliable forms of birth control have been available for all this time. Still, the United States has bowed to pressure from political groups not to assist other countries with population control. The world's most populous countries, India and China, have not been successful in reducing their population growth. Although China has greatly lowered its population growth rate with its policy of one child per couple, the population of China may still grow by at least 8 million people. Human population growth is technically the simplest problem facing our Earth to solve. We have made concrete progress (lowering human population growth rates in the past half decade), but still we have a growing population because of social and psychological barriers to attaining replacement fertility in concert with increased longevity.

The ability to control the physical and biological world to ensure survival and comfort is an aspect of human nature that will not change. The best we can do is make food and housing available with technology that has less environmental impact. Most people will not go cold and hungry by choice, but they may be willing to choose technologies that are less damaging to the environment, assuming that such technologies are available and cost only a moderate amount more than existing technologies. Education about impacts of resource use is necessary before people choose to limit their excessive resource use.

The desire to display accumulation and consumption of resources is deeply ingrained in human society and seems to have been fashionable over much of human history. Display of accumulation and consumption of resources is probably partly a sexually derived characteristic. It is possible that display of wealth can be replaced with some other status symbol. Cigarettes have been considered sexually attractive and sophisticated at times, but in the United States and other countries, people seem to be changing their minds. Furs were a status symbol in the United States in the past, displayed by female sex symbols in movies. Now animal fur clothing is not nearly as fashionable in some parts of society as it used to be. One can imagine it being out of fashion (against the social norm) to be a cheater who uses too much of Earth's resources to improve status, but that is probably too optimistic. Education about the true impacts of human activities on our global environment is necessary before individuals and societies alter their behaviors. This education may also limit hoarding behavior.

Cooperative behavior exists in an "us against them" context. Countless people have argued for a society with global cooperation in which all people on Earth have the same opportunities and rights, with no discrimination and no poverty. Few people would argue with the goal of an egalitarian global society, but we certainly have not attained it. If humans cannot cooperate with one another and treat them respectfully, what chance is there that they will cooperate to protect the environment until it is viewed as in the best interests of each individual and each separate country? Again, proper education is paramount to encouraging cooperative behavior, but education alone will not solve global environmental problems.

Cheating (using more than a fair share of resources) is a behavior that a proportion of society will always display. Penalties must be made and enforced or incentives instituted so that cheaters view costs as too great relative to the benefits derived from cheating. Since this cost analysis by cheaters includes determining the benefits of not cheating, incentives for environmental protection will help promote global protection. Internalizing environmental costs of economic activities, providing incentives for those who have less environmental impact, and imposing substantial penalties on those who do not pay external costs will all be needed to modify global human behaviors. Given the selfish behavior of people, voluntary internalization will not occur.

Game theory as applied to human behavior demonstrates that people are willing to punish cheaters at a personal cost. In the real world, these costs may include providing incentives (a cost to those paying for the incentives) or creating permit or emission credit trading systems (also some cost for administering and enforcing these programs). However, the public will not be willing to assume these costs unless they are aware (educated) that a portion of the society is using more than its fair share of Earth's resources. A social norm of acceptable environmental behavior will need to be established, and an educated public is key to establishing a social norm. Accurate accounting of the value of Earth's resources will be necessary to convince the public that it is in the best interest of humanity as a whole to cooperate with schemes to regulate global environmental impacts.

Changing our behaviors so that society inhabits the Earth in a sustainable fashion will be a tremendous undertaking. The behaviors outlined in this volume will be altered only with a concerted effort from society. Technological, political, and personal changes will be required to alter the current trajectory on which we find ourselves. Will the needed changes come from science, religion, politics, or some other sphere of human society?

Where Will the Solution Come From?

Some of the areas society has traditionally turned to for solutions to problems have so far failed to provide solutions to global environmental impacts. Human endeavors that could contribute to solutions are science, religion, social science (particularly economics and behavioral science), environmental activism, corporate policies, and political action. These are the questions: Can science provide a "magic bullet"? Can religion change people's behavior? Can an economic solution be devised? Can social scientists devise behavioral solutions? Can historians find solutions from past societies? Can corporations curtail their expansion of developing and using resources? Can political groups agree to cooperate on maintaining the environment in which we live? Can environmental groups provide the leadership required to nudge society in appropriate directions?

Will "Hard" Science Provide a "Magic Bullet"?

Some technological optimists do suggest that science can solve all our environmental problems, as well as deal with limitations on how many humans the Earth can support. But most of us realize that science may not be able to provide easy answers to how to use our environment sustainably for several reasons, including the unwillingness of society to employ existing scientific advances that could improve our efficiency of resource use and decrease population growth and the lack of commitment to solving global problems through science. Scientists behave like other people and mostly do only what they are given support to do. Science may not provide a magic answer to our looming environmental problems, but we already have much scientific knowledge that could be used to protect our environment. So, in one sense, the question that serves as the title of this section is not inclusive enough.

Society does not put all existing scientific information to use for a variety of reasons. We know how to conserve energy, minimize pollutants, control human fertility, manage many diseases, and feed all people on Earth. But governments often do not implement policies required to accomplish these goals. For example, cultural, religious, and political pressure often leads to a lack of support for population-control programs, even though it is scientifically possible to provide reliable contraception worldwide. Some in the energy industry use the political system to avoid initiatives to increase energy efficiency, and many industrial and agricultural lobbyists work against efforts to minimize pollutants.

Even worse, governments may support environmentally harmful policies under the guise of environmental protection. A potential example of "environmental" policies gone bad is the promotion of ethanol (gasohol) as an environmentally beneficial fuel. As early as 1980, it was realized that using all the corn grown in the United States to produce ethanol as fuel would supply only about 7 percent of the energy consumed in the United States, and this did not even account for the energy required to grow the corn.[12] By 2004, energy use rates were such that our entire corn crop would produce only enough ethanol to equal 2 percent of the annual oil used in the United States. Accounting for minimal estimates of the energy required to grow the corn reduces the rate to about two-thirds of 1 percent. Growing crops to produce ethanol requires fertilizer and machinery, and producing ethanol from crops requires transportation and energy for refining. One study estimates that 1.29 times as much energy is used to produce a gallon of ethanol than the ethanol will yield when it is used as fuel.[13] If it requires more energy to generate ethanol than it creates, more greenhouse gases and other environmental damage are produced than by just burning the same amount of oil. But this accounting is controversial, and the issue is unresolved.

Currently, the very best yields of ethanol from biofuels are about 1100 gallons per acre per year (1000 L per ha per yr). At this rate of ethanol production, it would take 10 times as much land as we currently have in cropland on Earth to provide the global energy used in fossil fuels. How do we feed people if all cropland is producing fuel? There are certainly places on Earth where biofuels are practical (in countries where petroleum supplies are mostly imported and population density is relatively low, as is per capita energy use). Biofuels are not a solution to global energy use, however.

Some studies suggest that the net energy yield of ethanol is positive (that the energy it takes to produce each gallon of ethanol from corn is less than the energy the ethanol will yield) but that producing ethanol from crops can lead to as much greenhouse gas emission as burning fossil fuels to provide the same amount of energy. These problems are not generally taken into account when economic benefits of ethanol are considered.[14] Using leaves and stalks in addition to the corn yields more energy than the cost of fossil fuels for production.[15] However, maintaining corn stalks in the fields after harvesting stops erosion, and removing them would exacerbate sediment pollution and loss of valuable topsoil. These calculations do not consider the environmental costs of fertilizer, sediment, and pesticide runoff. Also, the tremendous amount of water needed to grow corn, as well as to refine ethanol from corn, is not accounted for by most proponents of producing fuel from corn. Given the uncertainties of net benefit, it seems strange to push the use of edible crops such as corn and soybeans for energy production in a world where over 1 billion people lack proper nourishment.

Scientists have warned repeatedly of problems resulting from an increasing consumption of resources and a larger human population. The National Academies of Science of 73 nations have warned of problems associated with resource use, population growth, and global environmental damage and the need to make a transition to a sustainable world.[16] The Union of Concerned Scientists released the "World Scientists' Warning to Humanity" in 1992, signed by 1700 scientists, including most living Nobel laureates in sciences.[17] This declaration briefly mentioned most of the issues covered in the first few chapters of this book. Warnings about greenhouse gas releases have had little influence, however. Bill McKibben published a detailed account of greenhouse warming in 1989 that was widely read and commented on, but it spurred little concrete action.[18] The Union of Concerned Scientists' warning was followed up with a "Call for Action" in 1997, set forth at the Kyoto Climate Summit and signed by 1500 scientists, including 110 Nobel laureates and 60 U.S. National Medal of Science winners, urging effective controls on human practices affecting climate.[19] Despite clear, scientifically accurate warnings for over 15 years, the rate of release of greenhouse gases has only increased.

Another reason science cannot necessarily solve our problems is that it does not have the solution to all problems. Or, at least, science is not always able to provide answers to problems when we need them. For example, huge amounts of money and effort have gone into curing the many types of cancer, but only moderate progress has been made. AIDS cannot yet be cured despite decades of significant research. Likewise, a clean source of energy that is cheap and easily available would go a long way toward easing global environmental impacts. Although curing AIDS or providing cheap, abundant, and environmentally benign energy would lead to wealth and fame for some scientists (there was tremendous initial excitement and press over false claims that cold fusion reactions could supply unlimited energy), no such advances have been made.

Technological advances often contain their own environmental costs. When automobiles were first introduced, they were touted as a solution to the health and aesthetic problems of horse feces filling city streets. The impact of air pollution from autos (locally and globally) was unforeseen. Likewise, it has been suggested that a hydrogen economy could decrease greenhouse emissions but lead to the destruction of upper-atmosphere ozone (endangering all life with increases in UV radiation). What appears to be a clean energy source could have its downsides as well.

Another reason science cannot provide answers is that not all scientists and engineers care about the environmental implications of their research. Designing a luxury car or sport utility vehicle is not going to decrease global warming, yet many are dedicated to monetary gains derived from professions in science and engineering.

An adequate food supply may be another problem science cannot solve. One triumph of modern science has been the green revolution, or the drastic expansion of crop yields that has occurred over the past 40 years. Without technological advances in this arena, we could not support all the people on Earth today. As Earth's population continues to grow and the standard of living increases, a similar expansion of food production will be required over and above what has already been accomplished. But food production per unit area of cropland has inherent limits. For example, per capita cereal grain production has not increased over the past five years.[20] With plant breeding, optimum light, nutrients, and temperature, crops can grow only so fast, given the physical and chemical laws of our universe. In addition, as discussed in chapter 3, the green revolution has come with a substantial environmental cost. Most of the green revolution is driven by the use of fossil fuels, chemicals (fertilizers, pesticides, and irrigation), and water. A clean source of unlimited energy could be invented that would allow us to artificially light huge buildings containing rack upon rack of food plants, or that energy could be used to convert atmospheric CO_2 directly into food without the intermediary plants, but I would not bet that this will occur soon.

Can science provide technology to support an expanding human population, grow more food per unit area of cropland, and control environmental impacts associated with the greater crop production? Although technological optimists say yes, we are gambling with the well-being of humankind by assuming that science will be able to solve all problems as they arise. The stakes could not be much higher, and, as the old saying goes, you should look before you leap.

Scientists can find only those answers they are funded to find. Thus even though the United States should be less dependent on oil as an energy source for economic, political, and environmental security, an extremely limited amount of funding is earmarked for research on alternative energy sources. If grants are available, scientists will apply for funds and might eventually find some answers. Take my promotion of support for research with a grain of salt, given the principle I put forth in the preface: the more someone has to gain from a statement, the less you can believe it. Because I am an environmental researcher, I have something to gain from promoting more funding for my area of research. Also keep in mind that scientists are likely to tell the truth about their research subjects regardless of consequences, because presenting all the facts is the ethical cornerstone of our profession.

Although scientists may never find a "magic bullet," we can learn many things about preserving Earth. Our scientific understanding of the Earth that supports us is incomplete. Scientists once attempted to support humans in a sustainable sealed system called the Biosphere, but our knowledge of plants, microbes, and animals is not extensive enough to create even such a small self-supporting ecosystem.[21] Those responsible could not completely seal the Biosphere and sustain human life for more than a few weeks without inputs of oxygen. Animal diversity could not be maintained in the small Biosphere rainforest; one species of ant dominated all other insects, and all animals other than humans had to be removed. The inability to control even a relatively simple ecosystem does not mean that incompetent scientists were running the Biosphere; they just did not have sufficient knowledge to make it work. How can we hope to have a predictive understanding of the capacity of a highly complex and diverse Earth to withstand the insults we visit upon it when we cannot even assemble the components of nature into a small isolated environment well enough to support human life? The precautionary principle seems applicable in this case.

We are only just beginning to understand the rules of thumb to be used in ecology with regard to species invasions and extinctions.[22]

Likewise, global biology and climate are characterized in only a rudimentary fashion.[23] Without considerably more funding for research on Earth's ecosystems, such an understanding seems unlikely in the near future. As need grows, so does scientific expertise. So why has environmental science not kept pace with global environmental problems?

Science may not provide a solution to environmental problems because scientists (like other people) are most concerned about problems that will immediately influence them and their chances for success. Most scientists do not study environmental problems. The developed world puts tremendous amounts of resources into the science of how to cure cancer and heart disease, and a large number of the world's top scientists are concerned with these and related problems. These diseases are experienced by both the people who study them and their relatives. Funding agencies and scientists will truly start to work on Earth's environmental systems and human behavior toward them only when there is funding to do so. Funding will come only when there is a general consensus that Earth's capacity to sustain our standard of living, including the health of individuals in developed countries, is in immediate peril. Until that time, issues thought to be more important in the short term will take priority.

Scientists want to be recognized for their work. The most prestigious awards in science are Nobel Prizes. Many of these awards go to scientists who make advances that improve the health of people in developed countries or create technological advances that raise the standard of living of people who already have more gadgets than they need to survive. No Nobel Prize for environmental research exists, although there are several prestigious, but less well known awards such as the Crafoord Prize, the Swedish "Nobel" for ecologists.

Another way to gauge what the scientific community views as important is by the types of science published in premier scientific journals. The two most prestigious, broad-topic, international scientific journals are generally agreed to be *Nature* and *Science*. I perused the first 100 research articles published in each of these journals in 2006. Molecular, cellular, and health biology were the most common subjects of the articles in these journals (47 percent) and are most often the subjects of press releases and awards. Small-scale physics and chemistry, particularly materials science, were well represented (20 percent), as were space science (15 percent), particularly planetary exploration. All the rest of the sciences—including geology, evolution, sociology, and paleontology—are covered in the remaining 18 percent. Ecology and global environmental change research reports make up about one-half of that 18 percent. Environmental contributions are not prominent in the most prestigious journals. Emphasis on areas other than environmental research must affect specialties chosen by new investigators; most promising young scientists notice which fields receive most attention and which fields are most prestigious.

Scientists as a group might be expected to be more aware of the scientific basis of our environmental problems than the general population. They have the training to understand scientific principles behind the greenhouse effect, ozone depletion, and other forms of pollution. Even scientists who are well versed in the environmental impacts of humans behave in ways that are destructive to our environment. I will use myself as an example. Obviously, I think that Earth is under pressure from human activities. Still, I have a high standard of living and use more than my share of resources simply by living as a middle-class U.S. citizen. I recycle some trash, ride my bicycle to work (about 37 times more energy efficient than a midsize car and better for my waistline), and turn the heat down in the winter. But I use more resources than necessary. I switch on an air conditioner in the summer; eat fresh vegetables throughout the year; use a car for most trips longer than a couple miles; fly on airplanes for pleasure trips; ride ski lifts; take long, hot showers; and undertake many other activities that use more energy and resources than are essential. Like most scientists, I know that my work does not directly address the most important issues in our world's future but only nibbles at the edges.

Can Religion Rise to the Challenge of a Global Environmental Crisis?

Religion has no built-in way to protect the environment. No major world religion has a *central* tenet that the environment should be protected. Religion is primarily concerned with human behavior toward other humans (morality) and God (or at least ultimate meaning), although some secondary support for environmental responsibility can be found in the world's mainstream religions.

In the United States, religion, particularly biblical literalism, is significantly negatively correlated with support for the environment.[24] People who believe in God are less likely to support spending money to protect the environment: a more rigid view of religion is correlated with lower support for environmental protection. Andrew Greeley, the author of the study that found these correlations, hypothesizes that biblical religion per se was not responsible for these survey results; rather, biblical literalists have a rigid, general political view that includes an antienvironmental sentiment.

The primary reasons for Muslims, Jews, and Christians to preserve our environment come from the concept that Earth was created to sustain and be cared for by humans.[25] Buddhism is concerned with the idea of interdependence of all things, and this can be interpreted as an environmentalist position.[26] Hindu writings describe the destruction of the world through overpopulation and environmental devastation by humans.[27] The concept of ahimsa, respected by both Buddhists and Hindus, requires that a person not kill any animal life. Chinese and African religions can also be interpreted to include aspects compatible with environmental protection.[28] However, most of these religions, and even the indigenous traditions in many parts of the world that are commended for their reverence of nature, do not directly address overpopulation and excessive consumption of resources, because all the major religions arose before these were serious ecological problems.[29]

In the worst case, some use religion to justify population growth and environmental damage. Some claim that God put us on Earth to multiply and made the world for humans to use however they please. Others believe that the end of this Earth is near—that environmental destruction is a sign of the Apocalypse—so how we treat it is of little consequence.

Environmental damage is common in societies dominated by all religions, even those in which the government is a theocracy. A strong component of environmental protection in all the world's major religions would probably not solve our environmental problems; a substantial proportion of people ignore religious ideals. For example, most major religions preach against material desires and accumulation of wealth, but few people heed these ideals.[30] As already discussed, it takes only a portion of the world's population using resources at greater than sustainable rates to seriously deplete resources for all.

The idea that Native American and non-Western religions have an innate respect for the environment has been suggested as a possible avenue for ethical protection of the environment.[31] Unfortunately, societies with naturalistic religions are not immune to destruction of the environment around them. I have discussed how the Anasazi and Mayan civilizations may have collapsed in part because of nonsustainable use of their environments. Reverence for nature does not ensure that it will not be overexploited.

Religion is an example of human behavior that gives hope of modifying actions in spite of basic human nature. Major religions promote many unselfish behaviors. They are able to survive and attract followers even though many concepts they promote are contrary to selfish behaviors innate to humans (although they may play to self-interest, such as an ultimate entrance into heaven). Thus a religion-based movement to preserve the global environment could help transcend basic human desires to use more resources.

Any solution involving human behavior as it influences the global environment must include religion, because most people adhere to a religion. The Gallup-International Millennium assessment conducted in 60 countries showed that two-thirds of the people in the world feel that God is an important part of their life. Religion guides many people's views of right and wrong. A big step would be instilling a moral obligation in people now to protect the environment for future generations.[32]

Some religions are moving toward promotion of environmental stewardship.[33] The International Parliament of World Religions, attended by 8000 people in 1993, issued the statement "Global Ethics of Cooperation of Religions on Human and Environmental Issues."[34] International meetings on the environment and religion were held in 1988 (Oxford), 1990 (Moscow), 1992 (Rio de Janeiro), and 1993 (Kyoto). In England, the Alliance of Religions and Conservation publishes books on religion and environment, and in the United States, the National Religious Partnership for the Environment (NRP) joins Jews and Christians concerned about conservation of the environment. These are small steps in the right direction.

Some religious leaders are realizing the ethical importance of preserving our environment. In 2002, Pope John Paul II and Bartholomew I, patriarch of the Eastern Orthodox Church, issued their "Common Declaration on Environmental Ethics," stating that environmental degradation was "betraying the mandate God has given us: to be stewards called to collaborate with God in watching over creation in holiness and wisdom."[35] Reverend Richard Cizik, vice president for governmental affairs of the National Association of Evangelicals, has said that "polluters will have to answer to God, not just government."[36]

Religions do not have to be moribund. For their own survival, and that of all people, religious leaders must respond to the challenge of helping to solve modern environmental problems. Will the world's religions emerge as leaders in the morality of environmental protection? For the sake of humanity, I hope so.

Economics

The ability to manage our socioeconomic circumstances, and regulate the relationship between environment and global economic activity, will be central for global environmental management. Economists have not yet provided strategies, or the strategies have not been adopted by policy makers, to stabilize and equalize economic activity around the world. In many ways, economists are similar to ecologists (although they do have their own Nobel Prize). Just as ecologists could not maintain the Biosphere, economists do not understand everything about economics. If they did, many of them would be fabulously rich.

The global economy is volatile and can change rapidly. Economics has not provided the ability to buffer the economy from perturbations. For example, when terrorists attacked the World Trade Center and the Pentagon on September 11, 2001, the economy of the United States and the rest of the world suffered to a much greater degree than the direct economic damage caused by the attacks. Long-term stability would require a less-fragile economic situation than we currently have, and economists will need to point the way to attaining stability.

Historically, little recognition of the economic importance of environmental factors has taken place. Many economists treat environmental considerations as "externalities"; they are difficult to quantify, so they simply are not considered in economic equations. Many conservative economists do not believe that excessive human use of our planet's resources will constrain human economies and well-being. These economists think that people will substitute other resources when they run out, and that human ingenuity will solve any environmental problems that arise. The ability of Earth to support humans indefinitely, regardless of what they do to it, has been taken for granted by many economists. But some economists do not take the environment for granted.

Researchers in environmental and resource economics recently began attempting to quantify the monetary importance of the services that ecosystems provide to humans. These approaches value the materials that ecosystems produce (such as fish from the ocean and trees from the forest) and the services they provide (such as removal of disease-causing organisms from drinking water by self-purification), and they examine how much people are willing to pay for the use of unpolluted habitats for recreational, cultural, or aesthetic reasons.[37] Additional surveys can determine how willing people are to accept environmental damage and how unwilling they are to accept risks in the face of the unknown. Psychological research suggests that people are less likely to accept risks of factors they have less control over (such as environmental toxins) than they are of those that they control (such as alcohol consumption).[38] People also seem to be unwilling to accept some forms of symbolic environmental damage independent of pure economic value. For instance, when people realized that the use of DDT was leading to the extinction of the U.S. national bird, the bald eagle, many were unwilling to accept this and pushed politicians to limit use of DDT.

One example of the worth of ecological conservation is illustrated by a study reporting that land prices are substantially higher for property bordering lakes with higher clarity.[39] This analysis looked at increased sales prices of property near lakes and conducted environmental measurements to establish water clarity, and then determined that there is direct economic value to maintaining water quality in these lakes.

A preliminary economic valuation of Earth's ecosystem services by Robert Costanza (one of the pioneers of the new field of ecological economics) and his colleagues put the yearly global benefit of ecosystems at $33 trillion in 1997, and this value has probably increased since then. The value of Earth's ecosystems is related in large part to values of nutrient cycling, raw materials, cultural benefits, waste treatment, water supply and regulation, and atmospheric gas regulation.[40] While some economists question the specific estimates made by Costanza,[41] the idea of economic valuation of the environment has become a vigorous area of economic research.[42] The following discussion of some of these values illustrates the approach.

Nutrient cycling has value because the process maintains water quality. Raw materials (for example, lumber, hay, and fish) are provided by natural ecosystems and have a clearly defined market value. Cultural values are strong: few people find destroying the air quality of the Grand Canyon acceptable, few deny that the Great Barrier Reef should be protected from harmful nutrient runoff, and few condone wholesale destruction of tropical rainforests and the diversity they contain. More locally, indigenous people depend on the preservation of natural environments for the very survival of their cultures. Whaling is a central part of Inuit culture, but if there are no whales, the tradition will disappear.

Waste treatment is important for stopping the spread of human disease into water supplies. Modern sewage-treatment plants compress the natural cleansing ability of rivers and streams into a series of ponds and holding tanks where harmful bacteria and organic waste are removed. In lesser-developed countries, untreated sewage enters lakes, rivers, and marine environments. As long as population is not too densely packed, these habitats can self-purify. Rivers and estuaries near many large cities in developed countries have already had their capacity for self-purification exceeded.

Water supply and regulation is a greatly underappreciated value of ecosystems. In forested areas, precipitation falls and moves slowly into streams. If an area is deforested, rainwater rushes across land, collecting sediment and other pollutants, and floods through streams and rivers. In Costa Rica, several reservoirs were built for drinking water supply and flood control. Drainage areas above reservoirs were then logged. Within a few years, the reservoirs filled with sediments and were unable to fulfill their function. This was a direct demonstration of the economic value of forest preservation and ultimately led to the creation of forest preserves.

In 2005, tropical storm Jeanne killed over 2500 Haitians with flooding and mudslides from their deforested watersheds. Only 1 percent of the original forest remains on the Haitian half of the island of Hispaniola. In the Dominican Republic, the other half of the island, much natural forest remains, and flooding and mortality were substantially less from the same storm. Economic and direct human consequences of deforestation related to water supply and regulation are clear in this instance.

The regulation of atmospheric gases was discussed previously. Forests sequester greenhouse gases. A significant portion of increases in CO_2 that are causing global warming is related to deforestation. In the future, countries that maintain their forests may receive economic compensation from others that want to release greenhouse gases. Following a concept called "emission credit trading," each country would receive an allowance and could transfer that allowance for a designated amount of pollution release.[43] A country could release a smaller amount of greenhouse gases or even encourage absorption of the gases (say, by reforestation) and receive economic credit for the emissions. This program assigns a direct economic value to a specific ecosystem service.

Wetlands are the most valuable habitat per unit area because they purify water, provide habitat for wildlife, and protect surrounding land from floods. Coastal wetlands are nurseries in which many species of marine fish reproduce and spend the early part of their lives. Since wetlands are at the interface between land and many freshwater habitats, they intercept a lot of pollutants as they leave land and before they can contaminate water downstream. Many migratory waterfowl rely on wetlands' bountiful production of insects and small crustaceans to fuel their transcontinental migrations. This type of valuation of habitats has led many people who once thought wetlands were useless, mucky swamps that needed to be filled in, to recognize that the preservation of wetlands is important. Over the past few decades, protection of wetlands has been emphasized, in part because of explicit recognition of the economic and cultural values these habitats provide.

Costanza and his co-workers claim that their methodology results in conservative estimates of the worth of global ecosystems. The true value of Earth's services may exceed even their impressive estimates and substantially surpass the gross world product. An independent analysis placed the benefit-to-cost ratio of conserving our remaining wild natural lands at 100 to 1.[44] As the remaining wild lands become even rarer, their value will increase. Ecosystem goods and services undeniably have value, and as the economic practice of valuation continues to develop, it will become more effective and better represent the monetary worth of ecological goods and services.[45]

Economic approaches will be required to protect our global environmental resources; if people realize that ecosystem services have real monetary value, they are more likely to push their governments to help preserve ecosystems. Economists who disregard the importance of global environment to human economic systems are doomed to make poor long-term macroeconomic forecasts. Fortunately, some economists are beginning to study the interface between humans and their natural environment. The Association of Environmental and Resource Economists has 800 members, the Environmental Protection Agency has established the National Center for Environmental Economics, and the broadly interdisciplinary International Society for Ecological Economics has 1000 members.

The absence of a formal structure to account for environmental externalities in traditional economics is a major oversight. Accounting for ecosystem services moves economists toward the ability to quantify externalities. These accounting methods provide the first glimpse into how much corporations and individuals that destroy ecological capital should compensate society. We have already witnessed a change in the way some industries behave. In the past, it was standard practice to dispose of industrial wastes without regard for the long-term consequences to human populations. Societal response to incidents such as the pollution of Love Canal, in upstate New York, for which industry was held economically responsible for human suffering caused by chemicals dumped into the environment, is a first step toward a society that demands accounting for environmental externalities.

Game theory, external costs, unregulated resources, and unequal economic payoffs must be considered when formulating international, national, and even local environmental policy.[46] Economists are at the forefront of game theory and predicting certain aspects of human behaviors related to economics. As discussed in chapter 5, game theory can be directly related to global environmental issues. Some economists are paying attention to environmental issues and ways to maintain sustainability. When an economist wins a Nobel Prize for an environmental application, it will be clear that the field of economics as a whole takes environmental issues seriously.

Sociology, Cultural Anthropology, and Psychology:
An Important Piece of the Puzzle

Although natural scientists and engineers could offer technological advances and economists could craft effective economic incentives to help change the way humans treat the global environment, social sciences will provide an indispensable part of cultural change.[47] Sciences such as sociology, cultural anthropology, and psychology are less predictive in the sense that the future behavior of large groups of people is the most difficult variable to forecast accurately because of the numerous known and unknown interacting variables. Behaviorists, however, can provide appropriate models for successful social programs.[48] For example, scientists and government agencies can supply new crops considered to be more nutritious, but sometimes people will not eat them; they prefer their traditional foods. An anthropologist can establish the cultural biases that are likely to make a specific program successful or unsuccessful.[49] Understanding the human cultural, behavioral, and political dimensions will be necessary to educate people and provide acceptable ways to protect the global environment.

Social scientists have been working on the behavioral aspects of environmental issues. The Society for the Psychological Study of Social Issues was founded in the 1930s. Members of this group have studied the psychological aspects of environmental issues.[50] Funding for such studies has been meager, and only recently has study turned to global environmental issues. Still, this is a promising new subdiscipline of psychology that is emerging to directly deal with issues of conservation and environmental protection.

Social scientists need to make the jump from behavior of people in small groups to that of a global society. Research on game theory indicates that people are more likely to cooperate in small groups when there is communication among players and when moral appeals are made before game play.[51] As a result, maybe using small group–based solutions to large-scale problems is required. One route to small group–based decisions may be to stress the management of resources on the community level, such as control over local land use and water supply.[52] The extensive research on the tragedy of the commons indicates that a number of conditions can lead to avoidance of the tragedy.[53] Unfortunately, these conditions generally do not apply to global environmental problems. Scaling up to global institutions and common resource pools is one of the major challenges facing social scientists.

At a much more basic level of human behavior, social neuroscience is approaching the biological aspects of how we interact as a social species, and combining this approach with traditional psychological methods may yield insights into how large groups of people interact.[54] We are just starting to understand what motivates people interacting one on one through game theory experiments and other psychological tests. Scaling up will be the next step.

Lessons from History and Archaeology

Civilizations have succeeded or collapsed in the past. We are doomed to repeat previous mistakes if we do not study their causes. In the past, the fall of a civilization often meant that another would sprout up somewhere else nearby. These collapses were often a direct or an indirect result of environmental overexploitation. If we overexploit our *global* environment, there unfortunately is nowhere else to go. Lessons of history and culture merit careful attention.

Jared Diamond is a modern-day renaissance thinker who has substantially influenced the way people view the world with respect to history, culture, and environment. Simultaneously with his active physiology research program and as a medical school faculty member, Diamond developed a career as an ecological scientist based on his work on the birds of New Guinea. He speaks a dozen languages and is a Pulitzer Prize–winning author for his analysis of culture in the book *Guns, Germs, and Steel*. There is perhaps no more qualified scholar to analyze the links between human culture and environmental damage.

Diamond hypothesizes that there are five reasons that civilizations collapse: (1) damage societies cause to their environment, (2) climate change that affects the ability of a society to support itself, (3) hostile neighbors, (4) loss of cooperation with neighbors (as opposed to direct attacks), and (5) political, economic, social, and cultural issues.[55] His analyses of a number of case studies suggest that usually several of these reasons coincide to cause the collapse of civilizations. The two of Diamond's points most relevant to this book are that environmental damage was an important problem for societies in the many cases he studied, and societies that failed catastrophically could not adapt socially and preserve the environment that supported them. Although Diamond provides ample support for his hypotheses, historians and anthropologists will continue to refine the approach he describes and perhaps find ways that past societies consistently avoided disaster.

One example Diamond discusses in detail that illustrates some of the choices we face as a global society is the collapse of the Greenland Norwegian society. The colony was successful before 1300 because the climate was relatively mild. When the Little Ice Age started, the climate naturally cooled and it became more difficult for grazing livestock to survive. Norwegians in Greenland overused their grazing land as its resources diminished. Simultaneously, support from Norway declined, and political issues made survival even more difficult. In contrast, the Inuit people living in the region at the same time survived on available foods such as fish. Archaeological evidence (lack of fish bones in the Norwegians' trash) suggests that the Norwegians were unwilling to eat fish. Thus the inability of the Norwegians to survive in Greenland was not because it was impossible to live off the land but because they could not do so solely on beef and dairy products. Greenland Norwegians had contact with the Inuit, so they could have learned survival techniques from them. What cultural conditions caused Norwegians to not adopt the ways of the Inuit and improve their chances of survival in the face of environmental challenges? The breakdown was rapid enough that Greenland Norwegians knew it was happening, yet they were unable to change their culture to adopt Inuit survival strategies.

As a global society, we now recognize the dangers of our global environmental impacts on Earth. We also know many of the steps necessary to protect the planet. Will humans in today's societies take the necessary actions to adopt sustainable cultural practices? Perhaps anthropological, historical, economic, and social-behavioral researchers can find answers to these questions, in part by studying past societal collapses and by studying the interplay among human culture, environment, economics, resource use, and long-term survival.

Capitalism Must Be Regulated in a Sustainable World

Capitalism has spread around the globe, and environmental impacts are driven by what people do in the context of capitalism. The success of capitalism is one strong piece of evidence for the idea that humans behave selfishly. The entire system is built on people acting in their own self-interest. The failure of Communism, a system that relies on social cooperation, further exemplifies the fundamental self-interested behavior of most humans. The assumption that people would behave altruistically for the greater good is challenged by the collapse of the Soviet Union and the conversion of China to a more capitalistic society. Capitalism currently runs on growth, increasing development, and consumption of resources.

Peter Barnes is the cofounder of Working Assets, a very large and successful financial company based on the concept of socially responsible investing. He wrote a thoughtful analysis of the relationship between capitalism and the environment.[56] Barnes points out that any company that cannot continue to generate short-term profits will fail, and the law requires managers of corporations to attempt to increase the value of corporate stock. He documents how international corporations have ever more power and influence. Sales by Fortune 500 companies accounted for one-third of the U.S. domestic product in 1955 and two-thirds in 2004. Barnes also details how corporate profits can be used to fund lobbying and influence legislation, including legislation related to the environment. The basic characteristics of capitalism have led to the question: Is sustainable capitalism possible?[57] This question is difficult to answer, but I hope the answer is yes. The challenge is determining how to reach sustainability in a world that is increasingly capitalistic.

Some argue that corporations should be asked to implement only voluntary measures for environmental protection. Self-policing by corporations is putting the fox in charge of the hen house. The inability of many individual companies to self-regulate is why antitrust laws are absolutely necessary to maintain a healthy economy. I am not aware of an example of a large corporation making a major environmental concession without being forced to (either by direct public pressure or, more commonly, by government regulation). In some cases, environmental concessions are profitable, or at least very inexpensive, and are readily adopted. Most large corporations fight environmental regulations that might lead to loss of profits with every means possible. They are bound by law to do what they can within the law to increase profits, including making inaccurate scientific arguments and lobbying politicians who run the world's countries.

The debate over controlling sulfur and nitrogen emissions that result in acid rain exemplifies how industry can overestimate costs for compliance with pollution-control regulations and then adjust to the regulation far more efficiently than it initially predicted.[58] In 1982, energy-production groups predicted that regulation would cost them $2800 to $6400 per ton of sulfur-emissions control, while the U.S. Office of Technology Assessment predicted a $130 to $195 cost per ton. A few years later, an independent economic assessment of real costs established a true value of about $150 per ton. An obvious disparity existed between estimates made by the group of companies that were claiming massive loss of profits and those of the less-biased auditors.

The corporate sector of society has provided numerous beneficial goods and services for people around the world, but mostly without regard to environmental impacts such as pollution and rapid consumption of finite resources. Any effort to control the global consequences of human population growth and resource use must consider corporate behavior, which probably is more predictably selfish than the behavior of any other societal group. Companies will usually attempt to maximize short-term over long-term profits. This is why economists emphasize that incentives (such as short-term benefits for not polluting) may be more effective than punishment (the threat of eventual long-term cost).[59]

Maximizing profits will be pursued at the expense of the well-being of individuals, governments, and the environment. But history has verified that, even with regulatory and societal constraints, many corporations will operate successfully (for example, the tobacco industry). Thus corporations will almost always argue against regulation that will affect profits but will also usually find the most efficient way to operate under the constraints of environmental regulation.

Interestingly, the insurance industry is currently the largest financial group concerned about greenhouse gases and global warming. Insurance companies are worried because global warming brings more weather extremes (for example, more devastating hurricanes, cyclones, and other storms). Their concern reflects the potential for losing substantial amounts of profit from paying claims if the number and intensity of "natural disasters" increases, however; it is not necessarily an ethical approach to business.

Some corporations behave in an environmentally accountable manner because it is in their financial best interest, and a few do so because their owners think environmental responsibility is an ethical responsibility. A recent example is the push by General Electric to cut greenhouse emissions to 1 percent below their 2004 levels by 2012 (even if all corporations met this goal, it would not slow global warming because it targets only 2004 levels).[60] General Electric is taking this step because extensive research into its customer base led the company to believe that it would be beneficial. Entire marketing strategies are being built on creating an image of "green" responsibility and the possibility of taking advantage of the growing number of consumers who make decisions based on environmental considerations.[61]

Examples of green corporate behavior build optimism in that they prove corporations can pursue profit with less damage to the environment. Unfortunately, these corporations are exceptions. Unless "pro-environment" cooperators make up a large proportion of the total population, they will be the exception and their behavior will not drive global impacts. The general public can compel corporations to act in an environmentally benign fashion. If a company loses sales or contracts due to its environmental impact (real or perceived), it will be forced to change or face collapse. Consumer habits, public perception, and government controls guide how corporations will behave. A public that is educated about the externalities of corporate behavior and willing to work with companies to protect our environment is the first step toward reconciling capitalism with long-term sustainability.

Two examples illustrate how informational programs can encourage market-based solutions to environmental problems.[62] One is enactment of product-labeling requirements. The U.S. Department of Energy developed the "Energy Star" label, marking products that are relatively energy efficient. Some evidence indicates that these labels make consumers more sensitive to energy price changes. Europe and Canada have similar labeling programs. The second example is the enactment of reporting requirements. If firms are required to publicly report the pollution they release, the public may be more likely to demand that emissions decrease.

I am not aware of any examples of environmental regulations putting previously healthy major corporations out of business. Environmental regulation does not have to cause harm to the general economy or a net long-term loss of jobs. The economy actually can shift to create jobs for those involved in pollution control. While environmental protection may not stimulate the economy, it may not cost as much as some would have us believe. Furthermore, long-term economic stability depends on long-term environmental stability. Corporations concerned primarily with short-term gain usually will behave in a less environmentally responsible fashion than those with long-term goals.

Why Democracy and Other Political Systems Are Ineffective at Protecting the Environment

The very nature of political institutions and politicians make them unlikely to solve problems without a substantial and sustained popular push for change. Democracy is spreading throughout the world. But democratic and all other major political systems currently controlling the world's nations allow substantial amounts of environmental damage.

People often vote selfishly. Politicians act over the short term to get reelected. Environmental issues are long term, and the general public does not currently understand the economic benefits of environmental protection. A politician who wants to get elected is wiser to convince voters that he is the candidate for economic improvement than to promise that he will work toward a cleaner environment. The typical line we hear from politicians is "Of course I am in favor of protecting the environment, but ... ," which means that other issues take precedence over protecting the environment.

In spite of what people say, they often vote for whom or what they perceive will enhance their own economic position. I find it inconceivable that people of modern democratic societies will vote for policies that will seriously curtail their resource use. No politician in the present-day United States could ever get elected with a platform of less driving, less electricity use, and fewer luxuries for her constituents. However, people do not always vote to maximize their economic well-being.[63] Civil rights has been an issue that has won politicians' elections, and all who voted for those politicians did not stand to gain directly from increased civil rights. Cultural evolution trumped self-interest in this case. I hope that cultural evolution will trump short-term self-interest in the politics of the global environment.

Nations that are not democratic are no better at environmental protection. For example, China has deferred environmental protection in favor of economic development.[64] This postponement has led to tremendous stresses on the nation's environment, including horrendous pollution, chemical spills, serious overgrazing, and cultivating marginal lands. Ultimately, China will be the key country determining global environmental impacts as the Chinese increase their per capita resource use.[65] A benzene spill in the Songhua River in 2005 rendered river water unfit for drinking for months, with the problem moving into Russia as water flowed downstream. Overgrazing and cropping marginal lands has increased desertification, leading to loss of vegetation and massive dust storms. These storms create health problems in Korea and are transporting dust around the world.[66] The dust even eventually impairs visibility in the Grand Canyon, half a world away.

To make the possibility of a political solution to our global environmental problems even more remote, entities that want to extract resources for profit have a large amount of money. Money is political power. Thus even if a political process for making environmental protection decisions exists, people who want to curb such protections can pay professionals to testify against such controls, pay scientists to cast doubt on solid research conducted by other scientists, and donate large sums of money to political campaigns. On the other side, environmental organizations operate on donations and volunteer labor. Although they do not directly make money from protecting the biosphere, they have to convince people to make donations. Thus environmental organizations generally are disadvantaged when it comes to essential means of achieving effective political influence.

Environmentalism to Date Has Mainly Been About Cutting Losses

Environmental groups are almost always fighting a losing battle. The question is not usually to what extent can we restore damaged systems, but, rather, how much of what remains should be protected. Just because one battle to protect a forest from being chopped down is won does not mean a related battle will not need to be fought again in the next year. Thus environmental activist groups have to be satisfied by minimizing losses, and they almost never gain ground (restore habitat that has already been damaged).

The greater the pressure is to exploit any ecological resource, the more difficult it will be to win battles to protect the environment. For example, logging companies have cut down most of the largest trees in the United States. Stewards of the few remaining old-growth forests are under intense political pressure to harvest the trees. Even if these forests are protected for a decade, some emergency (a wood shortage, a war?) in ten years could make it seem imperative to cut the forests down.

The most biologically diverse forests on Earth are in the lowlands of Indonesia. These forests were protected 20 years ago, but recent social and political unrest is making enforcement of protection difficult. Currently, these "protected" forests are subject to uncontrolled logging by companies that will profit at the expense of many species' extinctions. Indonesian forests will be gone by 2010 at current rates of logging.[67] Indonesian deforestation is an example of how environmental protection depends on societal stability. If global environmental damage reaches a point that it destabilizes global economic systems and societies, problems will never be solved, regardless of the actions of environmental groups.

One of the most important contributions that environmental groups can bring to solving global problems is to facilitate education. They can serve as the conduit of information between scientists and the general public. It is a rough world for survival of environmental organizations, but they have become adept at getting their message across, at least well enough to ensure their own survival. This need to survive, unfortunately, can lead organizations to overstate particular cases and exaggerate some environmental dangers. In the long run, these over-statements harm the environmental movement. The second arena in which environmental groups can be useful is to propose sound policy solutions. The ability to advocate successfully with limited resources in a system with high-paid corporate lobbyists is somewhat dubious.

What Basic Change Will It Take to Alter Our Course?

All people would benefit in the long run from protecting Earth. Political and economic instability is inevitable when resources become severely limited. Sustainability requires stability, which will occur only in a just society in which people do not starve or die prematurely simply because they have the misfortune of being born in a country with few available resources. Long-term stability should be the goal of individuals, companies, countries, organized religions, and global political organizations. If we wait to start a global movement toward stability, bringing human society to a point where it can live sustainably will become more difficult and the probability will increase that we will cross a threshold of catastrophic environmental damage from which we cannot return. The fact that humans (and their institutions) innately behave in ways that ultimately damage our global environment will make it exceedingly difficult to reverse behaviors predominant in the industrialized world.

People allow their political systems to guide drastic alterations to society under dire circumstances. Until now, this has occurred mainly in times of war. Under the pressure of war, people are willing to substantially alter their resource use rates, withstand loss of some freedoms, and not expect rapid monetary return on work toward solving a problem. A "war" to save the environment is one possible way the political process could dramatically alter human behavioral patterns that are leading to global environmental damage.

The recent "war on terrorism" in the United States illustrates the changes in behavior that people are willing to make when an outside threat is perceived to exist. Unfortunately, terrorism still exists, the "war" on drugs has not seriously controlled illegal drug use, and the "war" on poverty did not erase poverty in the United States. These examples suggest that a "war" model may help make headway over short periods of time, but people probably will not operate in crisis mode over the many decades necessary to solve problems like poverty or global environmental degradation. A "war" approach may serve as an initial trigger to substantial change, but more fundamental and dedicated change is required for long-term solutions to global environmental problems.

Al Gore has proposed an environmental "Marshall Plan" inspired by the success of many nations rebuilding Europe after World War II.[68] His plan has six goals:

1. Stabilize world population.
2. Create and develop environmentally friendly technologies.
3. Create more realistic accounting that assigns value to the environment.
4. Negotiate and approve a new generation of international environmental agreements.
5. Educate the world's citizens about the global environment.
6. Establish the social and political conditions most conducive to sustainable societies.

The problem is convincing people to implement such a sweeping plan.

We are called to assist the Earth to heal her wounds and in the process heal our own—indeed, to embrace the whole creation in all its diversity, beauty and wonder. This will happen if we see the need to revive our sense of belonging to a larger family of life, with which we have shared our evolutionary process.

In the course of history, there comes a time when humanity is called to shift to a new level of consciousness, to reach a higher moral ground. A time when we have to shed our fear and give hope to each other. That time is now.

—Wangari Matthai

No More Business as Usual

Transcendence, Enlightenment, Rationalization, Hope, and Action

A quote from Woody Allen provides a caricature of environmentalists from those who would prefer to ignore our problems: "More than at any other time in history, mankind faces a crossroads. One path leads to despair and utter hopelessness. The other, to total extinction. Let us pray we have the wisdom to choose correctly."[1]

I prefer the inspirational view of Nobel Prize winner Wangari Matthai, as expressed in her acceptance speech for the Nobel Peace Prize. Matthai is Kenya's most celebrated environmentalist and an advocate for democracy and human rights. She established the Green Belt Movement, which encourages farmers (mostly women) to plant and nourish tree seedlings. These trees create green belts that prevent soil erosion, provide shade and firewood, recover wildlife habitat, and improve the standard of living. Perhaps more important, the Green Belt Movement has empowered people to take control of their environment through local action to help improve community stability. Matthai put her idealism to work and changed the Kenyan view of environmental destruction; her vision is spreading through Africa. Our challenge is to expand her vision to a global scale and prevent society from destroying the planet that supports it.

Jared Diamond suggests that when trying to solve problems, humans can fail in their decision-making process in four ways. First, they may not anticipate a problem before it arrives. Given the momentum of global environmental problems, this type of failure may be particularly disastrous. Second, people may not perceive a problem when it does arrive. This book and many other efforts at environmental education are aimed directly at this second failure. Third, people may not try to solve a problem once it is perceived. Fourth, it may not be possible to solve a problem, even though people try. The third and fourth of these failures will be avoided by humanity on a global scale only if we can transcend basic characteristics of group human behavior.[2]

We must create a situation in which a great enough proportion of the world's population transcends basic genetic and current cultural drives to attain global sustainability. Now is the time to choose our path, and I am hopeful that we will make the correct decision. To transcend our basic drives and modify our behavior, we need to accept intellectually the necessity of doing so. This need for transcendence has led others to note that preserving Earth and its capacity to support us and other species is basically a moral decision.[3]

Environmental enlightenment is defined here as learning about the full implications of our global environmental impacts. It includes understanding momentum and that global environmental problems will not be solved immediately. Enlightenment will be necessary to influence most of humanity to behave in ways that overcome hardwired genetic propensities or ingrained, culturally determined behaviors in order to make our world a better place. An ability to drastically alter behavioral patterns contrary to our evolutionary program is what sets us apart from all other animals. A general definition of enlightenment is education resulting in understanding and the spread of knowledge. Here I focus on global environmental enlightenment. The Enlightenment was an intellectual approach during the late eighteenth century that led, among other things, to modern scientific thought.[4] Now we need a second enlightenment leading to a sustainable planet.

Enlightenment could allow human society as a whole to realize that global environmental problems threaten our long-term survival as a species. Enlightenment is necessary because until now, no natural selection has acted on humans to monitor global-scale environmental issues or to plan for global sustainability for future generations. Transcending our evolutionary and cultural basis is necessary to control impacts on our global environment.

One approach to this form of environmental enlightenment is "The Earth Charter."[5] This document was created after the United Nations World Commission on Environment and Development, in 1987, called for a statement that would set forth fundamental principles for sustainable development. After 13 years of consultation with thousands of people and hundreds of organizations across the world, the charter was finalized. It sets out the basic principles necessary to establish a just and sustainable world. Putting such principles into effect will require finding ways to get most people to behave contrary to their cultural and biological conditioning.

How Can We Transcend Human Nature?

We can use science and rational thought to perceive our world in ways that were not available to humans when natural selection and early cultural evolution was refining who we are. Our new capacity for perception of the globe opens up a possibility to change quickly in response to diffuse and distant stimuli. Thought and culture can change much more rapidly than human populations can evolve genetically. The ability to radically transform thought and culture has enabled humans to become the dominant species on Earth. Many animals have a hardwired fear of fire and will do anything to avoid being near flames. Humans overcame the fear of fire and harnessed a vital tool. I hope the ability of humans to change thought and culture can be used to ensure our survival and preservation of the environment we rely on.

Changes in thought and cultural evolution have led to collective human behaviors (social norms) that are not beneficial to each individual acting in a completely selfish manner. For example, robbery may benefit an individual and his family (kin, as discussed in chapter 6), but stable societies generally attempt to control it because uncontrolled robbery would destroy society.

Enlightenment can allow us to see beyond the short-term immediacy of situations and act over the long term for the best interest of all of humanity. Enlightenment allows us to understand that it would be dangerous if individuals had unrestricted rights to kill one another. The concept that murder is wrong is a human societal construct that allows people to coexist without killing one another. This concept may even have genetic roots.[6] Thus most people think murder is wrong and a person is bad if he commits murder, even though there is no evolutionary reason not to murder an unrelated individual and many animals kill others of their own species without hesitation. Humans also have times (war, capital punishment) when murder is socially acceptable. But coexistence in a stable society requires general agreement that most murder is unacceptable and should not be allowed.

A system of behavior for the good of society forms part of behavioral codes. The ability to form the concept of what it is to be a "good person" and behave in ways that are contrary to our basic instinct is based in part on the ability of culture to transcend characteristics instilled in us by biological evolution. This enlightenment allows more social stability.

Moral codes embedded in religion were an early major step in societal enlightenment and provided a basic guide for behavior independent of our evolutionary program. Although some aspects of moral religious behavior may be driven by our evolutionary nature, many aspects seem to transcend simple self-interest.[7] Transcending human nature (narrowly defined as the pursuit of self-interest) has been the province of religion for much of recorded human cultural history. Religion can codify behaviors that allow peaceful and productive coexistence of people, even at the expense of short-term gain for the individual. But religion can offer some individual reward, such as eternal salvation, for forgoing short-term gain in society, and thus religion may be perceived by some to be beneficial to the individual.[8]

The Ten Commandments deem, among other things, "Thou shalt not covet thy neighbor's wife." The ability to spread genes has a definite evolutionary advantage, and adultery is a fact of life. The commandment is an example of societal enlightenment because it goes against basic human nature in an attempt to avoid the strife that occurs among people as a result of selfish behavior. Social stability is enhanced if human males curb their desire to spread their seed. Strife caused by individual males mating with others' wives probably had the greatest impact on the small, insulated groups that were more common earlier in human society (at the time the Ten Commandments were written). Nonetheless, religious codes of conduct can help to lead to more peaceful coexistence among people. Some cheat on religious codes, as game theory predicts. But if enough people follow the basic code of conduct (social norm) that governs society, the behavioral code is useful and self-propagating.

Technological enlightenment came more slowly, but, over the past few centuries, it has vastly changed relationships between humans and their environment. Nothing in our evolutionary program is a specific adaptation to drive cars, operate computers, use toasters, fly airplanes, or use any of the other modern technological wonders that are a part of many people's daily lives. Technological advances have improved many people's lives tremendously. Antibiotics can prevent a cut finger from leading to a fatal bacterial infection. Air conditioning makes living in hot climates tolerable and allows the ill or weak to survive conditions they might not otherwise. A reliable supply of clean and fresh water and food, made possible by technology, increases both human health and happiness. Television and radio provide entertainment to billions of people.

Technological advances have changed the way we view and interact with the world. Technology has simultaneously increased our environmental impact to global scales and given us the global-scale ability to detect these impacts. Neither is part of our basic evolutionary nature. The next dimension of environmental enlightenment could be behavioral and technological. This dimension could lead people to sustainably coexist with other species that inhabit our planet and preserve Earth's ability to support us by harnessing our behaviors and our technology.

People have realized for centuries that, in a just world, all individuals should have the basic necessities for living: food, shelter, and security. Not all people feel this way, but our societies have moved toward more explicitly caring for everyone, not just those fortunate enough to be born into upper classes or richer countries. Thus conditions can change through enlightenment, and what is generally considered to be right and wrong can be transformed. We now live in a world where basic human rights are considered by many nations to be something to which all people are entitled. While we have a long way to go toward the goal of providing a basic sustaining level of food, security, and self-determination for all people, these goals are widely recognized as desirable. Only recently in human history have human rights become an important part of international politics. The next step is to globally recognize a goal of sustainability and maintenance of Earth's environments.

For most of human evolutionary and cultural history, mitigating environmental impact has been a poor strategy; individuals in small groups who did not use resources at maximal rates reproduced less frequently than those who did. By the 1960s, perhaps for the first time in history, many people throughout the world realized that the human population was growing explosively and that efforts should be made to control it (although the concept had been around for 100 years, and some people still disagree with it). Across the globe, many countries recognized the need to decrease human population growth, even though they were unsure as to how. The governments of the most populous nations on Earth, India and China, began attempts to control their populations in the 1950s and 1960s. Despite their efforts, the populations of both countries are still growing. China has been somewhat successful at curbing growth (even with only 1.7 children per woman, however, its numbers continue to increase due to momentum), but programs in India did not work for the most part. The greater success in China than in India is probably attributable to the totalitarian Chinese government's ability to more directly influence human behavior, although the ethics of how such influence was exerted are questionable.

Strong controls are necessary to even slow population growth, given the hardwired desire of people to reproduce. Despite the existence of technology that allows us to control reproduction, global population growth rates have continued to increase for decades. Only in the past decade has the *rate* of increase of population leveled off (as explained in chapter 2, the population is still growing, just not quite as quickly). Thus although many countries understand that unlimited population growth is undesirable, and political moves have been made to manage it, control over human population growth is still very limited.

There are examples of local and global levels of environmental enlightenment. Industrialization led to dangerous levels of air and water pollution in developed countries. As people realized that pollutants could directly harm their health, laws were enacted to prevent further pollution. Industry first denied that problems existed and then claimed that it would be too expensive to stop polluting. Despite these arguments, and legal and political posturing by corporations and other entities to avoid regulation, developed countries passed laws decreasing the amount of bacteria-born diseases entering our water from sewage, the average amount of air pollution each car produces, the amounts of toxic compounds emitted by industry, and the amount of CFCs released into the atmosphere. Many other types of pollution, such as thermal pollution of streams and lakes by industry, have also been controlled. Now that developed countries have moderated some environmental impacts, how will developing countries fare?

Some claim that economies go through stages as they develop, causing unavoidable harm to the environment, but that economic development eventually protects the environment. This argument is based on the observation that some environmental improvements have occurred in developed countries with a relatively high standard of living, despite economic analyses predicting that such improvement will not be stable in the long term.[9] While it is true that developed countries have less of some types of pollution (cleaner water, less air pollution in cities), they are still creating a larger ecological footprint by dispersing the pollutants they generate.

If historical precedents apply, an unavoidable amount of pollution is required to increase and maintain a higher average standard of living in lesser-developed countries. Most people on Earth live in developing countries. Thus if economic development requires pollution, substantial global environmental damage will be necessary to bring the standard of living of those in developing countries up to that of the developed sector. This concept has been understood for some time; when asked if India would follow the British pattern of development, Gandhi reportedly replied, "It took Britain half the resources of the planet to achieve this prosperity. How many planets will a country like India require?"[10] Since most people live in developing countries, increasing their standard of living will incur much more damage than was required to bring up the standard of living in developed countries.

The basic premise that developed countries are less polluting than developing countries, based on the observation of more stringent pollution controls in developed countries, is not necessarily correct. While pollution levels may be less in developed countries, the picture changes when considered on a global scale. Environmentally damaging industries and businesses often move to developing countries with laxer laws against polluting (although other economic factors come into play as well, such as income disparities and wages, availability of raw materials, tax incentives, and transport costs).[11] In many economic cases, developed countries can subsidize clean air and water at home by allowing poorer people elsewhere to assume the burden of pollution. Developing countries will not have the luxury of polluting other countries (unless they find weaker countries that they can force to assume the environmental burden) as they try to catch up with the economic advances made in developed countries. In fact, many countries that are thought to be developing may be getting poorer if we account for the value of the loss of their natural ecosystems and resources. Considering ecosystem value accentuates the trend of rich countries continuing to get richer while poor countries get poorer.[12]

Despite the potential negative aspects of pollution control in developed countries (for example, moving polluting industries to poor nations), these nations do illustrate a change in the way pollution is viewed politically. In the United States and Europe in the 1950s, only a few scientists and political activists were claiming that pollution was damaging the quality of life, whereas most of the public did not think that the environment was a major concern or were simply unaware of pollution. During the 1960s and 1970s, impacts of pollution on the environment became great enough, and awareness increased enough, that a concerted political effort to control pollution began. Still, during the 1960s, politicians were successful at the national level claiming that protecting the environment was not important and that it did not need protecting. Now, however, politicians in Europe or the United States cannot get elected if they are completely against protecting the environment (instead, they say, "I am in favor of protecting the environment, but . . ."). People in many developing countries are also concerned about global environmental issues. Thus our global society has begun to become environmentally enlightened. Can that enlightenment extend to addressing global environmental issues and resource use?

Our global situation is like a runaway train with a monkey as brakeman. We do not know if the train is headed for danger or if it will eventually coast to a stop. The monkey brakeman has purely evolved capabilities; he is capable of pulling the brake but cannot necessarily comprehend the necessity for doing so. The train has so much momentum that danger down the track must be recognized long before it is reached. Our ability to use logic to reach the enlightenment that transcends our basic evolutionary traits sets us apart from other primates.

Three Global Environmental Threats with Different Outcomes

Three major global environmental threats are used as examples: (1) increased UV radiation in sunlight reaching Earth because of ozone depletion, (2) global nuclear war, and (3) the greenhouse effect. Although all three (particularly the first two) threaten human existence on the planet, they have been dealt with differently. Examining these threats can illustrate some potential human responses to global environmental problems.

Ozone depletion and subsequent control of CFC production and release is the first example of a global environmental problem that resulted in significant international action (chapter 3). This action probably will eventually control the problem, although it has taken a long time. Environmental protection occurred because people and their governments recognized an immediate threat to their existence. The debate over controlling CFCs included some people insisting that no harm was caused by CFC release and that even more scientific study was necessary before the production and use of CFCs could be curtailed. As is currently occurring with the greenhouse effect, scientists were found who were willing to argue that no harm would come from CFC release. The debate slowed international negotiations to limit CFC use, but agreement ultimately was reached.

In 1987, eleven countries representing two-thirds of total global CFC use signed the Montreal Protocol. The agreement was to freeze CFC use rates at 1986 levels with complete phase-out by 2000. Initially, several very populous countries—Russia, China, and Japan—did not sign the agreement; by now, all major countries have agreed to limit CFC production and use. In general, industry supported banning ozone-destroying CFCs only after companies developed compounds they could sell for more money.[13] But the bottom line was that a global environmental threat was perceived and action was taken to counter the threat. The size of the ozone hole has yet to decrease significantly as of 2007, so it is not yet clear if the action has had the desired effect.

In direct contrast to the response to the threat of environmental devastation from ozone destruction, global society has not made major progress toward removing the threat of nuclear destruction. The threat of nuclear war still exists a half century after it became a real possibility. The possibility of global environmental destruction triggered by atomic warfare is real because an explosion of a moderate number of nuclear devices could put enough dust into the air to plunge the world into a nuclear winter.[14] Cooling effects of dust injected into our atmosphere has been verified by the global cooling that occurs when large volcanic explosions spew dust high into the atmosphere.[15] If enough nuclear weapons were detonated, even if they did not directly kill everyone, the entire Earth would plunge into a cooling catastrophe, killing crops and destabilizing ecosystems. Even areas not directly affected by radioactive fallout would be devastated. All-out nuclear war would lead to the extinction of humans. Scientists and others have warned of these dangers for decades, and most people understand that it could occur;[16] yet we still have enough weapons of mass destruction armed to destroy all human life on Earth, and ever more countries are obtaining the capability to produce nuclear weapons.

It would be simple to remove the threat of destruction of all human life from nuclear war. Destroy enough weapons, and global nuclear war cannot happen. Despite this straightforward solution, the political power of the nuclear-military-industrial complex has persuaded the governments of the free world to continue the production and maintenance of vast nuclear arsenals. A number of nations have acquired (North Korea) or are striving to acquire (Iran) their own nuclear arsenals because of perceived threats from neighboring countries and for reasons of national pride. We are still perched on the edge of destruction. Somehow, people have come to accept this threat and go on with their lives. This ability to deal with a constant threat is most likely due to denial and rationalization (as discussed in the next section). A clear threat to all humans and our environment from nuclear arsenals is present. Why can we make global efforts to curb ozone depletion but not reach agreement on preventing global destruction from nuclear arms? Perhaps the difference in action is spurred by the distinction between a threat that is in the process of occurring (ozone depletion) as opposed to one that is not ongoing (nuclear war) and we can hope never will occur.

Increases in CO_2 and associated heating of Earth represent a global human influence that is already causing significant changes in the atmosphere and could have catastrophic results. It will take decades or centuries to reverse this effect. Cities in coastal areas will need to construct barriers to keep the ocean out, as sea level rises and destructive storms become more common. Additional areas will need to be opened up for agriculture because weather patterns will render some existing croplands unproductive. More extreme weather and higher temperatures should not exterminate humanity (unlike all-out nuclear war or dramatic increases in UV radiation, both of which would kill most of the plants and animals on which humans rely for their survival), but these changes could cause major damage to our economic and societal infrastructure.

Despite extensive impacts of the greenhouse effect and associated global change, humans are showing little inclination to decrease releases of greenhouse gases. World governments considered controlling release of greenhouse gases (the Kyoto Accord), but the only goal was to scale release rates to 1990 levels. The United States backed out of the accord and is unwilling to even cap CO_2 emissions at the current, environmentally unsustainable rates. The American public seems split as to what price it is willing to pay for curbing CO_2 emissions. On the one hand, the sales of sport utility vehicles and low-gas-mileage trucks and powerful automobiles remain strong in the face of rising gasoline costs. On the other hand, a recent poll of 1018 adults indicated that 59 percent favored an increased tax on gasoline to cut energy consumption and reduce global warming.[17] Given trends in energy use and CO_2 emission, atmospheric CO_2 concentrations will continue to rise. Arguments by politicians that high rates of energy use are tied to a healthy economy, the tremendous political clout of the energy industry, and the inability of the public to sway policy have combined to make real control of greenhouse gases to levels where they will not cause greenhouse warming unlikely in the near future.

Given the three threats (ozone depletion, nuclear war, and greenhouse gas emissions), reason for hope exists. Human society seems to have controlled one of these three threats. Still, the momentum of global trends must be kept in mind; global environmental problems are of such large scale and have so much drive that action must be taken immediately to mitigate impacts that will occur well into the future.

Rationalization and Other Mechanisms as Enemies of Enlightenment

Rationalization is the ability of humans to justify selfish behavior in the face of facts showing where the behavior will lead. We do things even though we should know better. Rationalization is essential for most humans to function in modern society. Most people believe they are basically good. They think that most of the time they do the "right" thing, which causes a logical conflict with the way they actually act. With regard to the environment, many public-opinion polls confirm that most U.S. citizens are concerned about the environment or consider themselves environmentalists.[18] This occurs in spite of the fact that Americans have among the greatest environmental impacts per person on the planet.

Sigmund Freud believed that humans were basically selfish individuals who act on unconscious motivations. Examples of selfish behavior on the global scale abound. Millions of people in the world die

181

and suffer each day from preventable diseases and malnutrition. Over 1 billion people live on less than $1 per day and do not have access to clean water.[19] The suffering goes on, while those of us living in developed, affluent society pay movie stars, athletes, CEOs, and the industries that support them tremendous amounts of money to entertain us. If this sounds like a superficial comment, let me be more specific.

The Bureau of Economic Analysis in the U.S. Department of Commerce estimated a gross domestic product of $46.5 billion for the U.S. motion picture and sound-recording industry in 2005.[20] At a cost of $27 billion per year spent on global health care, millions of lives could be saved in poor countries.[21] The United Nations World Food Program estimates that one in seven people go hungry in the world and that $1.7 billion is immediately needed to feed the world's starving people in emergencies, nourish pregnant women and children, and stimulate local food production. The World Food Program received annual international donations of $2.8 billion.[22] Global military expenditures are around $1 trillion, and less than 2 percent of this could fund the World Food Program. Incongruently, the U.S. government pays farmers not to plant crops and is encouraging the use of food (corn) to produce ethanol to fuel automobiles.

As a society, the United States moves 27 times more money through the motion picture industry than it would take for emergency food aid for the entire world. The movie industry is only one luxury. The video game industry is now more lucrative than movies. Manicured lawns, expensive vacations, recreational sports, swimming pools, expensive foods and goods, big cars, and many other things are pure luxuries that consume tremendous amounts of resources that could be used instead to help solve the world's problems.

People rationalize the treatment of the world's poor in a variety of ways, from blaming people without money for their plight because they are "lazy" to simply believing that individuals can have no discernable effect on the issue. A similar type of rationalization allows us to mistreat the environment that supports us without taking responsibility for our actions, because people generally care even less for nature than they do for other people. Not all people behave in ways that harm the Earth that supports us, but many do. Even with laws to protect dwindling resources, some will continue to abuse those resources, and rationalization will allow them to think they are not doing any significant harm, even if collectively they are ultimately destroying our planet.

The idea of a changing frame of reference exacerbates the effects of rationalizing environmentally destructive behaviors. If all that people have seen since they were young are polluted and degraded landscapes, they expect the same in the future. Since we are not evolved to detect long-term change, slow environmental degradation can also change perception. If you have not seen a child for a few years, it is shocking how much she changes as she grows up. Parents do not perceive this change unless they look back at old photographs. Likewise, if suburbanization causes degradation of a native landscape, it goes unnoticed except by those returning after many years' absence. Luckily, we have technology (data records, photographs, remote sensing) that allows us to perceive very slow and global-scale changes.

Rationalization may be easier to change than reproductive drives, controlling surroundings, acquisition and consumption of resources, limited cooperation, and a propensity to cheat. Once society views a behavior as bad, it becomes more difficult for individuals to rationalize it. For example, murder is more difficult to rationalize than cheating on taxes. Social pressure can change drastically and greatly influences what can be rationalized. Thus 100 years ago, companies that used rivers to dispose of untreated waste were not generally viewed as bad, but now egregious polluters are viewed as criminals. Absolute biological effects of pollution are the same, but perceived societal impact has changed, and people find it harder to rationalize pollution now than they did 100 years ago.

A number of additional defense mechanisms were proposed by Freud for dealing with unconscious motivations, and they are related to the human propensity to avoid protecting the environment even though people know they should.[23] *Intellectualization* is a thought process we use to convince ourselves that something is not our problem or that we are not the cause of it. *Displacement* occurs when we redirect our feelings from the object that elicited them to a different target. *Suppression* consists of putting the undesirable consequences of our unconsciously driven behavior out of our minds. *Repression* is the same as suppression, but it is unconscious. *Reaction formation* is a strong negative reaction (for example, attacking environmentalists' motivations). *Projection* is to see the problem in others but not in ourselves. *Sublimation* is when we channel our anxiety into productive behaviors. Finally, *denial* is used by many to avoid the existence of a problem. Education can be used to combat many of these defense mechanisms, particularly denial.

Denial is a common response to large problems. Some people, faced with the enormity of environmental issues, fear of consequences of global environmental change, and a perceived inability to alter human behaviors, simply deny that there are problems.[24] And even though humanity has been under the threat of nuclear annihilation for the past half century, most people do not worry excessively about that threat. On a smaller scale, social scientists have demonstrated that people who live immediately below a dam worry little about the dam bursting, those who live a moderate distance downstream worry the most, and then worry again decreases in those who live farther downstream.

Denial also plays a role in global environmental problems. When survey results of Swedish, British, Japanese, and American respondents were compared, Americans placed a lower priority on stopping global warming, and fewer believed that "global warming has been established as a serious problem and immediate action is necessary." The percentage of hard-core skeptics who do not believe there is problem is greater in the United States than elsewhere.[25] The country with the largest impact on greenhouse emissions has a population that is more likely to deny the effects of greenhouse gases.

Denial is in part related to the limited ability of people to accurately assess risks. Some risks are far overestimated, whereas others are far underestimated. For example, people underestimate their chances of dying from many common health problems (such as drug addiction, alcoholism, and lung cancer) and overestimate their chances of dying from others (tornadoes and botulism poisoning).[26] Our misperceptions may be based on incomplete information. I hope education will combat denial, and at some point the weight of evidence will be impossible to deny. Effective communication and education about human environmental impacts on Earth is essential to solving environmental problems. If most people do not even know that a problem exists, or deny it in the face of hard evidence, no impetus to solve it will develop.

Hope for the Future and Developing a Broader Perspective

Societies can be successful environmental stewards when the need to preserve our environment for future generations is recognized. Although I am not a social scientist, it seems to me that family farmers in my home state of Kansas who inherited their farms and hope to pass them to their children take much better care of the land than do the managers of corporate-run farms. Japan probably protected its forests because of the desire of its leaders to pass down a stable society to their heirs.[27] Living for the moment does not lead to preservation of the environment for future generations.

Humans respond to perceived short-term threats to their existence but can plan for the future. If a critical mass of people perceives that global environmental degradation poses a substantial risk to them and their descendents' continued existence, then they will probably modify their behavior in ways to solve the problem. Actions that societies can take in response to perceived threats are remarkable; the response to World War II is a case in point. Individuals on both sides of the war made massive sacrifices—from a lowered standard of living to sacrificing their lives. Incredible technological progress was made during World War II, including the development of antibiotics, rockets, and the atomic bomb. The response to the war illustrates that once humans decide to work on an issue and provide funds and personnel, they are capable of solving very difficult problems.

Will We Survive Current Global Environmental Trends?

I am moderately optimistic about human chances for survival in spite of the threat of nuclear holocaust or biological warfare. Ingenuity will help us persevere. But given current trends and existing technology, few people are likely to have as high a standard of living in the future as those in industrialized societies do now. Those of us who reside in developed countries are living in a golden age. We are using up our planet's resource capital at the expense of our children and grandchildren. The least we can do is appreciate the tremendous luxury we have. An analogy is spending your life savings on a one-night party and worrying about how much caviar the caterer provided. Never mind that the cost of the party is jeopardizing your children's ability to attend college, and you may not be able to pay your home mortgage or health care costs in the future. A scene in the Woody Allen movie *Stardust Memories*, in which Allen's character gets on a dirty train with a bunch of ragged, downtrodden people, is poignant. He watches a train full of rich, beautiful people at a party, drinking champagne, going the other way. The people in the undeveloped world must feel like this character when they watch television and movies from Hollywood. Most people in developed countries never appreciate the luxury of using resources at the rate we do.

Earth's population will continue to grow, people in developed countries will continue to use resources at current rates, and developing countries will increase their rates of resource use. This overuse of resources will continue until Earth as habitat reaches a breaking point. The pessimist in me thinks we will live in a world with few remaining wild animals, dominated by weedy species such as rats and cockroaches.[28] Zoos will soon contain most of the remaining large animals. In my darker moods, I predict that when Earth's resources become more limiting, social collapse will result, and even more environmental devastation will occur. In times of crisis, environmental protection is the first thing to go. When oil becomes scarcer and more expensive over the next 50 years, many will return to dirtier coal burning or nuclear power. Our air will get worse, and nuclear accidents and contamination will become even more prevalent than they are now. We will breathe foul air and live in a radioactive soup. Smaller coastal cities will disappear, and larger ones will build walls to hold back the ocean as global warming raises sea level. Most surface water and groundwater will be contaminated and undrinkable. Food will become a very valuable commodity as the limits of production are reached, cropland is at a premium, and human demand escalates. This increase in demand could lead to even more starvation than currently occurs, along with the attendant disease and suffering. I hope that time will prove me wrong, but we are doing this experiment on Earth while ignoring potential consequences.

People living in developed nations refuse to believe that their institutional structures are fragile. To understand how quickly our social and economic structure can be deeply impaired when an essential resource becomes even temporarily limited, one need only look at fuel shortages in the United States in the 1970s and in Britain in the late 1990s, the electricity shortages in California in 2000 to 2001, societal collapse in New Orleans after Hurricane Katrina, and global economic fallout after attacks on the World Trade Center. Even in societies that seem relatively stable, full-scale riots can break out when a popular sports team wins or loses a major competition. What will happen if widespread energy shortages occur? Given that all food, medicine, and water supplies depend on fossil fuels, how will fuel shortages impact society? How will political stability respond to regional shortages of water? Will terrorism increase as the gulf between the haves and the have-nots widens?[29] Will our political and economic systems survive the intense economic downturn that would result from severely limited resources? We cannot assume that they will.

Summary of the Book So Far

Numerous indicators show that, for the first time in history, we are having a global influence on Earth's ability to support humans and other organisms. Our environmental impact is expanding, as evidenced by accelerating increases in global population, energy use, economic activity, water use, greenhouse gas emissions, ozone depletion, deforestation, loss of cropland, species extinction, and species introduction. These trends all have tremendous momentum that will take decades or centuries to reverse, if they can be reversed at all.

Trends of increased environmental exploitation cannot continue indefinitely because we live in a finite world. Our current trajectory puts us on a collision course with the capacity of Earth to support humanity sustainably. This collision course is made harder to control by the fact that, almost without exception, environmental problems can be caused more quickly and cheaply than they can be reversed, and their scale gives them immense momentum. The United States spends billions of dollars cleaning up environmental contamination by industries that inflicted the damage to make millions of dollars in profits.

Basic human characteristics make it unlikely for most people to change their behavior. These characteristics include the drive to reproduce, the desire and ability to control biological and physical aspects of surroundings, the display of acquisition and consumption of resources in wild excess of immediate requirements, cooperative behavior only within defined groups and when it benefits the individual, and the maximization of individual gain. Even if all people do not behave selfishly, the remaining people who do drive exploitation of the environment and continue to expand humanity's footprint.

Tests of game theory confirm that a proportion of people will behave in an uncooperative fashion (cheat) but also suggest that humans are willing to reduce uncooperative behavior at a personal cost. The tragedy of the commons is the simplest environmental game and illustrates that any open-access resource will be overexploited by normal people acting in a rational fashion. Most global, open-access resources are being overexploited, including forests, water, capacity of our atmosphere to absorb pollutants, species, fisheries, and arable croplands.

Human behavior makes regulations governing resource use ineffective, because the proportion of cheaters is determined by the costs and benefits of using resources. Resources become more limited and more valuable the more they are used, so incentives to overuse them will increase. It will cost the public more to control the overuse by a smaller proportion of the population in the face of increased human population. The more people there are, the greater the absolute number exploiting the resource with a constant penalty levied by society for overexploiting a resource. Control may be in the form of pollution taxes, permit fees, regulations, incentives for moderate use rates, or other behavioral modifications.

Scientific, religious, economic, social, psychological, and political strategies are all required to solve global environmental problems. Substantial impediments prevent each of these spheres from providing solutions in isolation. Humanity finds itself in a place where natural limits will control our population and resource use rates if we do not do it ourselves. Due to the momentum imparted by global-scale resource use and population growth, it may be too late to fix problems once they are finally perceived.

Education is the first step toward providing the public with an appreciation for threats posed by indiscriminant resource use and population growth. Education may also improve the capacity of humans to appreciate the economic and aesthetic value of our Earth's biological systems. Without education, there is no hope that humans will control global environmental impacts, but education alone will not solve our problems. Accepting environmental education without rationalization or denial is the first step in the global enlightenment of humanity. Education alone is not sufficient; the United States is one of the most educated countries in the world yet uses resources at a greater per capita rate than any other.

Control of population growth and resource use rates by methods not of our choosing could lead to much suffering and could even cause a feedback of further destruction of Earth's ability to sustain humans, let alone our standard of living. So is there any chance to avert the collision between the immense momentum of humanity's propensity to destroy the environment that supports us and the finite nature of our planet? Can we manage our footprint?

Only in imagination can we travel . . . from the recognition of
environmental problems and the need for soundly based
policy; to the selection of solutions based on moral reasoning;
to the biological foundations of that reasoning; to a grasp of
social institutions as the products of biology, environment,
and history. And thence back to environmental policy.

—Edward O. Wilson

Consilience

Socioenvironmental Restoration and
Sustainable Inhabitation of Earth

Evidence presented in this book overwhelmingly illustrates that global
environmental impacts have momentum that puts us on a collision
course with Earth's ability to support us. The environmental catapult
imparts momentum that supersedes individuals, single nations, or a
generation of people. All major indicators of our global footprint—
human population, economic activity, water use, energy use, global
greenhouse gases, and global pollution—are increasing exponentially.

In chapter 1, I noted the contradiction between the following
observations:

1. The best predictor of future behavior of large groups of people is
   prior behavior: cultural trends have inertia in part because they are
   driven by repeatability of human behavior with deep, hardwired, and
   cultural roots.
2. Earth's support system is fragile, and its resources are finite.

These two observations are in direct conflict because we cannot maintain increasing rates of environmental pollution and resource use in a
finite world. Global trends have massive momentum; they are impossible to change in less than a decade and, in many cases, even in less than

a century. This momentum is also present in human behavior, as evolution dictates that our basic nature is not amenable to environmental preservation. So we are rapidly moving toward an impasse dictated by momentum and a finite Earth.

Humanity's current situation requires new solutions based on a variety of fields of study. Every category of knowledge we possess is required to overcome the large and complex problems of global environmental impacts. More recent trends in research and society offer promise for solving global environmental problems; all these trends are related to interaction of multiple threads of human culture and thought converging on the problem of controlling impacts on our global environment. Broad interdisciplinary approaches (consilience) are necessary to solve our problems, relevant fields of knowledge are becoming more interdisciplinary, and we must harness the tremendous intellectual capacity of humans to attack global environmental issues that we are enmeshed in.

Consilience: How to Live Up to Our Potential

Solutions to the tremendously difficult problems posed in this book require a new approach, because global environmental impact is a relatively new and unique problem for human society. Edward O. Wilson describes *consilience* as an integrative approach to solving problems by melding advances across disciplines.[1] Consilience requires bridging the gaps between arts and sciences. The discussion in chapter 8 makes it clear that no one discipline alone can solve our global environmental problems.

I propose a solution based on the convergence of many specialties plus the development of some tools and approaches that do not yet exist. I refer to the change in our society and its relationship to our global environment as "socioenvironmental restoration." The concept is to use modern sociological, ecological, economic, and other societal (for example, religious and political) tools to change our behavior and modify inhabitation of Earth toward sustainability. I describe some pieces of socioenvironmental restoration that are already in place and then introduce my proposal. Some areas where multidisciplinary approaches are making headway in solving environmental problems are environmental science; behavioral and social sciences; and economics, education, religion, and philosophy.

Revolution in Environmental Science as a Model for Consilience

The study of ways to improve our environment is making impressive gains in many areas. Science may not provide a "magic bullet," but at least it can offer knowledge of what will be necessary to catalyze global sustainability. Several areas of progress are very encouraging, and they involve forging cross-disciplinary approaches.

Ecology, as a scientific discipline, was traditionally limited to the study of organisms other than humans under fairly pristine conditions. The academic definition of ecology is quite different from the general public's understanding of the word "ecology": professional ecologists study how organisms interact with one another and their environment, whereas much of the general public defines ecology as environmentalism. In contrast to ecology, environmentalism is the belief in protecting the environment (an "ology" is a science, and an "ism" is a belief). Recently, more ecologists have started to consider the impacts of humans on ecological systems.[2] Wildlife managers have been doing this for years, but now the field of ecology is spawning subdisciplines that specifically include a more multidisciplinary approach. Examples of this include the fields of restoration ecology, urban ecology, conservation biology, and landscape ecology.

Restoration ecology is growing rapidly. Simple preservation is only part of the solution if we are to inhabit our planet in a sustainable and stable fashion.[3] If ecological damage exceeds a threshold such that ecosystem goods and services are no longer provided reliably, it is necessary to repair that ecosystem.[4] We need to learn to restore the function and diversity of many natural systems that have already been damaged. In an attempt to encourage ecosystem repairs, the Society for Ecological Restoration was founded in 1987 and has grown to more than 2000 members. In addition, most environmental agencies in most countries are involved in restoration as well as mitigation of existing impacts. Restoration ecology requires understanding the historical properties of natural ecosystems and how to restore and maintain those properties with continuous human pressures.

Urban ecology is in its infancy, but it is starting to provide information on the environment that is inhabited by more than one-half of the people on our planet. Urban ecologists must consider sociological patterns if they are to understand the ecosystems they study. For example, people who plant native vegetation may create very different ecological conditions than those who plant only nonnative vegetation.

In Phoenix, Arizona, people fertilize and water lawns and gardens in what was desert. This change in vegetation alters humidity and affects the other plants and animals in the city.

CONSERVATION BIOLOGY AND LANDSCAPE ECOLOGY

Conservation biology and landscape ecology have made tremendous strides in the past few decades and have emerged as separate disciplines. *Conservation biology* melds together genetics, ideas about how plant and animal populations move, knowledge of how organisms interact with one another and their environment, and public policy. This broad, cross-disciplinary approach is necessary to help predict what is necessary to conserve species. *Landscape ecology*, to quote from the International Association for Landscape Ecology, is "the study of spatial variation in landscapes at a variety of scales. It includes the biophysical and societal causes and consequences of landscape heterogeneity. Above all, it is broadly interdisciplinary."[5] Both these fields require sophisticated analytical capabilities (advanced mathematics and computer modeling), as well as an understanding of how humans interact with the environment. The science necessary to maintain rare species provides one example of how the fields of conservation and landscape biology incorporate disparate disciplines that would traditionally have been the province of separate specialists. Following is a specific example of how bridges across fields are necessary to conserve species.

Inbreeding depression occurs when a species is rare. Inbreeding depression can cause populations with too little genetic diversity to be doomed to extinction. Decreased long-term survival of species occurs because inbreeding with very similar members of the same species can unmask harmful characteristics. Humans from all cultures are well aware of this problem, and most societies have prohibitions against mating with siblings or first cousins. Likewise, zoos exchange animals to avoid inbreeding.

Conservation biologists have used mathematics, statistics, and computer modeling to predict how likely inbreeding depression is to precipitate the extinction of a natural population of endangered animals. In the real world, however, many populations are not completely isolated; rather, the range of most species becomes fragmented as humans destroy native habitats. Situations often occur in which rare species are found in restricted areas, and only a few individuals can move among these areas. But it is this movement that can counteract inbreeding depression. So, to conserve a species, people need to consider inbreeding in the context of fragmented habitats.

Landscape ecology considers how the arrangement of habitats over time and space influences species. In a very fragmented landscape, populations may not be connected genetically. If enough connection exists between the fragments to avoid inbreeding depression (that is, flow of genetic novelty is great enough to avoid inbreeding), then a population can survive indefinitely. This is exactly what zoos attempt when they exchange animals to mate; zoos are like very small fragmented islands, and they move animals to increase gene flow among islands and avoid inbreeding depression. Thus a long-term plan to ensure the survival of an endangered species would need to consider genetic diversity and the connections between suitable habitats (landscape pattern), as well as how humans may interact with the environment in the future.

USING NEW TECHNOLOGY TO SOLVE ENVIRONMENTAL PROBLEMS

The interdisciplinary natures of urban ecology, landscape ecology, and conservation biology exemplify part of the revolution in environmental science. Also, tremendous advances have been made possible by more powerful computers and their associated software. New statistical methods, mapping tools, and modeling capabilities allow us to deal with much more complex data than was possible previously.

The field of environmental science has recently seen an explosion of techniques and the ability to deal with complex data sets. State and national agencies, particularly in North America and western Europe, have collected data on environmental parameters such as water flow and quality, air quality, biodiversity, pollution emissions, and habitat destruction. In my research, human influences on water quality can be identified using data on land use and water chemistry that are publicly available. Such an analysis would have been difficult or impossible a few years back. Now, with the combination of computer power, satellite imagery documenting patterns of human land use, accumulation of water-quality monitoring data, and ability to combine data, new environmental recommendations can be made based on data analyzed by a moderately competent scientist on a regular personal computer. Our ability to predict what it takes to maintain Earth in such a way as to support humanity, even with expanded resource use and population, is rapidly increasing.

Over the past few decades, the ability of statistical tools to tease apart complex interrelationships among data has greatly increased. Just one new statistical technique, neural network modeling, uses computers to test a wide variety of possible networks of influence on an environmental factor of interest. This statistical technique mimics the way neurons in animals connect in networks and uses a model to "disentangle" complex interactions. Dozens of other statistical approaches are available for analyzing complex data sets and creating predictive models from those data. For example, environmental scientists may want to understand why an invasive species is successful in some habitats and not in others. Many factors are potentially important in complex ecological systems—such as light, temperature, moisture, wind, interactions with other species, and degree and type of human influences—rendering it difficult to dissect factors that allow invader establishment. Modeling can be used to rearrange linkages among factors into all possible combinations to provide the best prediction of success of the invasive species. If key factors can be determined, managers can then focus on and attempt to control them.

Neural network modeling and other statistical techniques require complex calculations that would take an individual with a calculator a lifetime to make. Today, environmental scientists can bring many statistical methods to bear on the same problem and find the one that works best, as well as compare results of one statistical model with results from alternative models.

Analytical tools for dealing with large amounts of data related to mapping the environment and linking these maps with other data have become much more powerful. Geographical information systems (GIS) enable us to make composite maps and analyze complex data across space. Remote sensing allows scientists to link images, such as those from satellites, with the maps. Spatial resolution of widely available satellite images is precise enough to show large individual trees. If you can afford it, you can hire an airplane to fly over an area and take a high-resolution digital photograph on which it is possible to distinguish individual shrubs and small trees.

Imagine a map with all the political lines drawn, and another over it with all the roads, and yet another with all the distributions of plants and how they change over time, and one with animals and how their dispersal and abundance change over time, and elevation and slope, and where water flows, and soil types, and croplands, and on and on. Each map is actually a layer with spatial coordinates, the exact latitude and longitude, precisely associated with each piece of data at a specific time. Any way that you can imagine these layers can be combined, they can. Exactly this type of map allows for tremendous power in predicting how human behaviors will alter the environments that support us.

Complex computer modeling of the global environment is another area in which environmental scientists have made tremendous strides. Predictions of the effects of global warming are based on models called "global climate models" or "general circulation models." A typical version of these models would divide Earth's surface into over 6000 rectangular areas covering the globe. Each rectangular area then has 20 different compartments layered above it to represent Earth's atmosphere at increasing altitudes. These models then calculate numerous variables for each point in time, such as land cover and vegetation's effects on climate, temperature forcing by the sun, energy dissipation into space, convection of heat and moisture, and cloud cover. Such models take into account millions of data points to make predictions across time.

The most powerful computers on Earth are used to run these models. For example, in 2003, Japan produced the Earth Simulator, then the most powerful computer. This computer is capable of a calculation rate of 40 T flops.[6] Flops are operations, such as addition, per second. One T flop is 1 trillion calculations per second. The computer is powerful enough to increase the resolution of global climate models from 40 to 4 square miles (roughly 100 to about 10 km^2). This resolution is fine enough to simulate phenomena such as hurricanes and how they are affected by global warming. Numerous research groups around the world have constructed related models, and results from these models are compared. The combined results of these models represent scientific consensus on predictions of the effects of global warming. Predictions of the models are now being verified by measurements of global environmental characteristics.

HOW INDIVIDUAL RESEARCHERS CAN APPROACH
LARGE-SCALE ENVIRONMENTAL PROBLEMS

Here is one small example from my own research of how new technology can expand our ability to deal with environmental data across larger areas. Keith Gido and I participate in the aquatic GAP analysis program, an initiative funded by the U.S. Geological Survey to catalogue what is known about species distributions in the United States.

The aim of GAP analysis is to identify the gaps in our knowledge (hence the name GAP). Our group, including some of our graduate students—Bob Oakes, Jeff Falke, and others—put together geographical data related to fish distributions in Kansas. Maps of land use—as represented by maps of the area's cropland, prairies, urban populations, forests, and vegetation lining stream channels—were obtained from government data sources. Maps of stream and river channels also were obtained from the government, and errors were identified and corrected. Fish data were collected from the Kansas Department of Wildlife and Parks, historic published records, museums around the state, and personal collection records. All these data were merged in order to use GIS to analyze them. Ten years ago, this would have been a difficult, if not an impossible, task.

The data were then analyzed with statistical techniques to predict the presence and absence of rare and endangered fishes, as well as noxious invasive species. Although this was the original intent of the project, additional uses were found for this newly linked data set. Records were analyzed to determine how human-constructed reservoirs in rivers influenced fish species found upstream of dams. Data were combined with water-quality records provided by the Kansas Department of Health and Environment to estimate how humans affect water quality in the region, and this information will potentially guide future water-quality regulations. Using the maps and water-quality data, we also predicted reference water quality in North American prairies before large-scale land modification. This result has practical implications for determining how agricultural pollution influences our aquatic ecosystems in the central United States, and how humans might alter their behavior to protect water quality. The next step is to involve social scientists (geographers and demographers) to relate human population shifts, extraction of water from the portion of the Ogallala Aquifer that underlies Kansas, and increasing agricultural practices to the changing water quality and fish diversity of the region. Our group is just entering this phase of research, but the opportunity to work across disciplines is exciting and promising.

This one small example, of research conducted on rivers and streams in Kansas, demonstrates how recent advances in linking data across disciplines and subsequent analyses, and melding ecology and social science, can help us understand human influences on the environment. If one small group of researchers can use these techniques to cover a fairly large area, it is easy to imagine how such approaches could be extended to the scale of continents and oceans. Now a continent-wide network of ecological observatories in the United States (NEON, or the National Ecological Observatory Network) will take us to the next level of ecological integration. In the field of ecological research, this is the equivalent of a supercollider, space station, or human-genome project. Although the ability to deal with large-scale environmental issues is increasing substantially, the problem is still how to use the information. It will obviously be necessary for environmental scientists to collaborate broadly to have an effect on public policy.[7]

Revolutions in Behavioral and Social Sciences

Behavioral and social scientists have increased knowledge of the human brain and how it works, the fundamentals of human behavior, how to assess human behavior, and ways to alter that behavior. Some areas, such as political polling and advertising, have developed tools for assessing and altering human behavior that could be used to protect our Earth instead of just selling products and politicians. Other areas, such as the study of innate human behavior, could assist in designing strategies to alter human behaviors to decrease the harm inflicted on Earth. Effective regulation will ultimately need to take into account results from some of the following areas of research.

THE HUMAN BRAIN AND MOTIVATIONS FOR HUMAN BEHAVIOR

Behavioral science has started to assess the degree to which behavior has a biological basis and which behaviors are more based on cultural or learned factors. For example, many types of mental illness can now be attributed to physiological or genetic causes. The ability to ascribe organic causes is tremendously helpful to clinical psychologists because they have a better idea of what can be treated and what problems they cannot hope to change with psychological therapy.

Today, specific behaviors can be linked to neurophysiological responses. For example, a recent study combined neural imaging (magnetic resonance imaging, or MRI) and studies of patients with lesions in their brains to examine how uncertainty affects decision making. Neural imaging allows researchers to see, with high accuracy, the regions of the brain that are active while decisions are being made. Patients with lesions have a disconnection between parts of their brains

197

that can result in unusual behaviors. In one study, researchers showed that uncertainty in games (the same types of games discussed in chapter 5) can lead people to behave differently than would be predicted based on what would logically be the best strategic decision.[8] Such experiments could be used to assign a degree of plasticity to behaviors and to determine what is not plastic because it is hardwired.

Game theory as studied by economists is now being used to predict social behavior.[9] Since economic decisions can have environmental ramifications, interaction with behavioral scientists may be fruitful. One way game theory is making advances is using computer models to explore sensitivity to basic "rules of the game" and how cultural evolution can lead to cooperation in games.[10]

There are clearly opportunities for economists, psychologists, and biologists to make progress in explaining the motivations behind human behaviors. For example, some economists subscribe to the idea that the economy evolves, so that evolutionary models from biology may provide some inspiration for economics.[11] Advances in understanding altruistic behavior as a natural outcome of selfish individual behavior from an evolutionary perspective are beginning to be included in economic theory.[12] Such broad understanding of the biological motivations behind human behavior will be required to control impacts of the entire human population on the global environment. Advertisers are masters of dissecting and altering human motivations and behaviors.

ADVERTISING, POLLING, SURVEYING, AND LINKS TO COST-BENEFIT ANALYSES

Socioenvironmental restoration will require general public acceptance, and successful advertising can influence public behavior effectively in aggregate. Advertisers take advantage of collective human psychological characteristics to sell their merchandise. To entice people to buy their products, they appeal to, among other things, the desire to appear attractive to potential mates, the threat of disaster (insurance), and the drive to increase social status. The influence of surrounding color on moods has even been used to determine store and package coloring.

Study of the psychology of advertising is a vigorous and extensive field.[13] For example, research has been conducted on the effectiveness of using political advertising that features either fear or feel-good messages to make a candidate memorable. This study examined how aspects such as color, music, and types of images affect perceptions created by messages.[14] Concrete images may spur peoples' estimation of risk, but shocking or graphic images may not be very effective in changing behavior. The graphic conveyance of information in realistic terms of the risks, along with communication of the probabilities that negative consequences will occur, may be the most reliable way to use advertising to alter behavior, including behaviors that lead to global environmental degradation.[15]

The public seems mostly unconcerned about the ethics of advertisers who are attempting to manipulate the purchasing behavior of consumers. Advertisements to alter harmful behaviors, including those trying to educate people about the dangers of smoking or drunk driving, are tolerated or even encouraged by society (regardless of their effectiveness). Thus it is possible that advertising associated with socio-environmental restoration will help change mass behavior as part of an effort to preserve the ability of our planet to support humans over the long term. It is possible for such advertising to be effective. For example, the slogan "Don't mess with Texas" was an anti-littering advertisement first; the slogan was so successful that it went beyond the original intent, but it was initially a highly effectively remembered message. Tools of advertising and polling are also based on multidisciplinary approaches, and new technology in these areas will be needed to educate the public about environmental implications of their behavior.

Polling and surveying to determine willingness to pay or public opinion are increasingly used in modern life. Willingness to pay is a key ingredient of assigning economic value to ecological goods and services. Politicians rely heavily on polling to understand the public's moods. Advertisers use polling and product testing to develop new products. Entertainment businesses, television in particular, use polling to determine popularity. These polls are driven by technological advances in communications, statistical advances that allow the determination of margins of error, social science to assess potential factors that may bias polling (for example, economic status), and behavioral science with experience to ask questions to extract relevant information.[16] Results of polling may be used to assign different levels of punishment or incentives to regulate environmental "cheaters." In an interesting twist from social scientists, incentives may be more effective if offered on a "lottery" basis. People are more likely to recycle if some occasionally receive a large award than if all receive a small, regular reward.[17]

In earlier chapters on behavior, I discussed how people weigh penalties relative to benefits to decide on a course of behavior. Management thus requires understanding the costs of penalties and the benefits of incentives for environmental preservation relative to the perceived benefits of environmental exploitation. Specifically, effective regulation requires determining how much people think preserving an environmental feature is worth relative to the perceived benefit of not preserving it. A successful regulation will make the perceived penalty for cheating or the benefits garnered by incentives greater than the benefit obtained by cheating. Polling could be used to determine when penalties for environmental damage or incentives not to cause damage are great enough to prevent most people from causing the damage. Society could combine game theory and collection of information on public sentiment to craft policies that could protect the environment. Polling could also determine the unwillingness to accept the risk that our environment may not be able to support us in the fashion we prefer.

ALTERING HUMAN BEHAVIOR

Finally, effective regulation requires understanding which types of behavioral modifications are successful. Psychologists have moved past the simple concept of laws and punishment to create a more complete picture of what is required to successfully alter human behavior:[18]

1. Education and persuasion are necessary, but they are not always effective.
2. A sense of choice and control increases participation. If people have a say in the regulations or systems used to control environmental behavior, they will be more likely to alter their behavior in a positive fashion.
3. Commitment is important. If people sign agreements to participate in environmental programs (energy conservation, recycling), they are more likely to continue their participation.
4. Information feedback improves participation in environmental programs. Programs such as gas mileage–information stickers on new automobiles, more frequent energy bills, and continuous readout of gas mileage on automobiles may lead to better conservation.

5. Peer pressure and social norms control behavior. As discussed earlier, people are more likely to comply with environmental regulations if others are obviously doing so as well.
6. Multifaceted approaches are probably required to facilitate lasting change.

Concentration on only one of the areas listed is less likely to be less successful than concentration on several or all of them. To concentrate on all these areas will require educators, social scientists, policy makers, and public participation. Artists, other creative professionals, and religious figures will be needed to help get the message across and create a cultural condition conducive to environmental protection. In general, lasting change is most likely if people understand it is in their own best self-interest.

Beginning an Economic and Second Industrial Revolution

Economics as a field has traditionally espoused economic growth as the ultimate goal for economic health. However, if we are to inhabit our Earth in a sustainable fashion, continuous growth is not necessarily a good thing. Growth to alleviate poverty and human suffering is easier to justify than growth simply to escalate luxury consumption by people who already live in comfort. As mentioned before, wealth is not a prerequisite to happiness.[19]

Ecological economics is a new transdisciplinary field that spans the social and natural sciences with the goal of obtaining a sustainable future. This is a revolution that is questioning the basic assumptions of classical economics—for example, growth as a goal and the lack of concern for equitable distribution of resources.[20] Ecological economics includes conventional economics, conventional ecology, environmental economics, and traditional resource economics.[21] The goal of ecological economics requires a new view of what makes for a healthy economy: long-term stability requires sustainability rather than growth to be the economic goal.

Steady-state economics is an alternative approach that advocates using technology to make economies more sustainable and equitable, not necessarily to encourage overall growth. Herman Daly was an early proponent and continues to be a powerful voice for steady-state economics.[22] He claims that traditional economics is very good at making predictions about economic markets with very restrictive assumptions, but many economists have retreated into narrow mathematical approaches. He thinks that economists should take the strengths of economics and apply them in a broader sense that considers scale (not just the local conditions of the market) and moral issues. Among other points, Daly and his co-workers are concerned about how the market can create conditions in which it destroys itself.

Global environmental damage to sustaining ecological capital caused by economically driven harm is a prime example of the potential for a market to destroy itself. This type of market failure is caused by not accounting for externalities. Externalities in this market failure are the values of ecosystem goods and services. Determining the values of these goods and services requires social scientists to assess perceived values and environmental scientists to calculate specific benefits related to different ecosystem states.[23] Consilience among economists, other social scientists, and environmental scientists is necessary to move a single field of study forward to help solve global environmental problems.

Another major area of economics that Daly and his collaborators are critical of is the theory of "discounting value." Discounting in the simplest case is a cost-benefit analysis. When a company decides if it should invest in a copper mine, it should do so only if it can make more money than it would by putting that money in the bank to earn interest. The problem is that discounting can be used to predict that resources should be used for immediate benefit because if not used immediately those that do not use them do not reap benefits (future generations may instead).[24] Discounting is one economic explanation for the tragedy of the commons. If cost-benefit accounting is used to assess environmental policy, the use of discounting by most mainstream economists is likely.

Under sustainable economics, the social system is assumed to remain indefinitely. Value for future generations is given equal weight to value for the existing generation. In this scenario, it is unnecessary to discount the value of resources saved for the future because society still has them. The Center for the Advancement of the Steady State Economy notes how the ecological footprint—the amount of resources used by each activity—tends to increase with economic growth as it is currently practiced. According to the United States Society for Ecological Economics, sustainable economics can be attained by finding the "optimal scale of the economy, efficiency in allocation of resources (goods and services), and the equitable distribution of this resource flow among ... people."[25] This society of economists exemplifies the potential revolution in economics that could help solve global environmental problems.

Economists concerned about environmental impacts are devising ways to assign values to ecosystem goods and services, including (1) determining the market value of goods and services produced by ecosystems, (2) evaluating ecosystem products or services that contribute to the production of marketed goods, (3) using pricing methods based on relative perceived values to assess how an ecosystem affects the price of another product (for example, the effect of a clear lake on property values), (4) determining economic activity generated by using an ecosystem for recreation (hotels, travel), (5) assessing contingent valuation or choice by questioning people about their willingness to pay for ecosystem value (such as conducting surveys of willingness to pay) or what tradeoffs they are willing to make, and (6) using established values for ecological benefits from one study and applying them to another (including transferring results from one region to another)[26] or assessing the cost of replicating the organization found in natural ecological communities.[27] Ecological valuation requires multidisciplinary approaches that include both environmental biologists and sociologists. Methods of economic valuation are strong enough to make management recommendations for specific areas.[28] Political attempts at ecological preservation also need to consider economics. Pollution taxes to discourage environmental damages are destined to fail if they are not designed with economic principles in mind.[29]

The new ecological economics (including steady-state economics, interfaces with ecological sciences, valuation of externalities, and consideration of human behaviors) will also need to consider the complexities of global society. Economic analyses of global environmental issues based on interactions of nations at the international level are being developed.[30] Such analyses will require consideration of international politics and reformulation of classical economics, yet another area where consilience is essential to solving global environmental problems.[31]

One issue that economists must deal with is that traditional measures of economic health do not necessarily account for other potential societal costs. For example, gross national product (GNP) is stimulated by environmental damage. The Valdez oil spill led to massive cleanup costs. These costs actually stimulated gross economic activity in Alaska over the short term, but the increase was not positive for the ecosystems of Alaska and the fishing and tourism that relied on them. Wars can also increase economic activity, but few would openly argue that wars are laudable. At least GNP could distinguish between national costs (for example, environmental cleanup) and national benefits (for example, production of valuable products).[32]

Some government subsidies to stimulate the economy are ecologically harmful. Examples include building an excessive number of roads, subsidizing marine fisheries, and driving the agricultural and forestry industries.[33] Full economic accounting shows that these perverse subsidies are counterproductive in the long run. New methods attempt to account for environmental resources in addition to estimating monetary flow.[34] These accounting methods require economists to collaborate with other social scientists, as well as with environmental scientists.

We can increase sustainability in many ways without harming our economy. In a fundamental sense, this requires economic approaches based on finite resource bases. Curiously, the field of microeconomics recognizes limitation of scale, but classical macroeconomists generally do not. Economic approaches that increase sustainability regardless of economic philosophy are also available. For example, we discard a tremendous amount of useful materials that could be reused or recycled. A new field has emerged that may help with this issue—industrial ecology. This field seeks to account for all the costs of discarding raw materials. Tons and tons of outdated computers are disposed of each year, and a full accounting reveals that it is economically disadvantageous to send them to landfills. These old computers have many parts that could be reused or recycled. If the energy required to make each computer part is included in calculating how much time and energy should be expended to recycle parts before a machine is discarded, reuse becomes far more desirable.[35]

At least some ecological economists are beginning to explore the possibility of placing an economic value on certain aspects of the environment that benefit humans. People are willing to go to great lengths to stop pollution in "their own backyard." If people recognize that the planet is "their backyard," then the potential economic value of preserving Earth increases substantially. Ultimately, a synergistic interaction among economists, behavioral scientists, and ecologists will be required to confirm the monetary value of maintaining Earth in a sustainable fashion.

Education, Religion, Philosophy, Architecture, and the Arts

Although biological evolution forms much of the basis of how people behave, education in formal or informal settings is also responsible for influencing how adults behave. The power of education to overcome innate biological self-interest is exemplified by the fact that many people are willing to die for religion. Religious terrorists do not necessarily commit suicide bombings because they are inherently evil; they do so because they have been educated to believe that it is the morally superior way to act. Suicide bombers likely feel they are righteous and religious, and they end up behaving contrary to their own self-interests.

Secular education can also alter human behavior. The U.S. Centers for Disease Control and Prevention reports that, between 1965 and 2003, smoking rates for adults fell from 44 to 22 percent.[36] This drop coincided with a large-scale public education campaign to increase awareness of the dangers of smoking. Will education help preserve the environment? Using education to help solve global environmental problems will require consilience among education providers, professionals concerned with ethical behavior, and those who study the consequences of behavior on our global environment. Assuming consilience, two educational phases are required. First, people need to be taught that preserving the global environment is the right thing to do for their own self-interest; second, they need information on what it will take to protect the environment and the consequences of not doing so.

In some cultures or sectors of cultures, religious education and other aspects of education are separated; in others, they are not. In cultures where religious education is the primary means of transmitting information to people, protecting the environment will require religions to teach that doing so is good or right. A revolution in how religious institutions view environmental protection will be necessary before religious education is an avenue to environmental protection.

Transmitting the idea that Earth should be protected and that such protection constitutes good behavior will be helpful even in cultures in which education is partly religious. In secular educational environments, philosophical ethical considerations form the foundation of right and wrong, and such ethics will need to include protecting our environment. These concepts of right and wrong must be carefully constructed to avoid contradicting the many possible religious backgrounds of students. Transmitting the idea that right and wrong includes protecting the environment is just the beginning of producing world citizens. Citizens with full environmental education will view sustainable treatment of the environment as a duty and living in an environment not excessively exploited by others as a right. Successfully transmitting the scientific, economic, and social concepts required to attain sustainability will help humanity reach the goal of stable coexistence with the environment that supports us.

Education is providing more information to more people than ever before. My children learned more at an earlier age than I did. Advanced topics 40 years ago—such as using computers, the genetic code, factors influencing global climate, and many other aspects of technology and science—are now common fare for preteens. Today's youth learn all this new information and still need to understand basic reading, writing, and arithmetic. Global affairs classes now have far more than just the newspapers, radio, and three television news channels the public relied on for information 30 years ago. Just as those who are adults now learned more at a younger age than their parents, our children must master more than we did. A complete education should include concepts of global environmental protection.

Education will continue to progress when it is combined with behavioral sciences. Simple theories of what constitutes a more effective educational strategy are no longer sufficient; rigorous testing of alternative methods using techniques established by social scientists is required to move the field of education forward. For example, beliefs and behavior are not always in sync with each other. General education may not be as effective as directed education for this reason. People have positive attitudes toward recycling that are more consistent with self-reported rates of recycling than are their attitudes of general concern for the environment. Still, education is important because even though people do not always act consistently with expressed beliefs, changing attitudes may in some cases alter behavior.[37]

Education will help change attitudes, but modern approaches are required. Experimental social scientists have established that feedback, modeling, and "framing" messages are more effective than just providing information. *Feedback*, rapid information on behavior, can help people understand the implications of their own behavior. For example, people save more energy in their homes if they are continuously fed information about the rate at which they are using energy. *Modeling*, providing people with examples of others' behavior, is also effective. Videos of a young couple showing how to conserve energy reduced energy use by similar couples, when compared with those who were not shown the video. *Framing,* using effective terms or examples, also improves environmental education. In one study, people were more likely to insulate their water heaters when told how much energy cost would be incurred if they did not insulate them, than if they were told how much energy they would save if they did insulate the heaters (the two concepts are the same but are "framed" differently).[38]

Additional features of effective education include getting people's attention, presenting credible sources, involving people in problems, and spreading information via people with social influence.[39] As with all the approaches discussed in this chapter, consilience (in this case quantitative psychologists working with policy makers and educators) will lead to more advances. Advances in the education system will make it easier to transmit information on the global environment.

Part of preserving Earth for anything other than selfish gain requires a fundamental belief that it is the correct thing to do. The first step toward engendering such belief is providing the public with information illustrating which people or institutions are consuming more than their share. The second step is planting the idea that it may be immoral to harm the environment. Religion is one major avenue (although not the only avenue to establish morals) toward convincing people that it is wrong to be selfish. Most religions are based on the idea that individuals must sacrifice to be good. The appropriate religious leaders could move humanity a long way toward acceptance of the idea that it is morally preferable to use only a modest amount of the world's resources for survival, shelter, and happiness.

Architects and builders have become involved in environmental issues by promoting "green" buildings. The American Institute of Architects has adopted a sustainable practices position that promotes the design of buildings that discourage greenhouse emissions through energy efficiency and other measures. The U.S. Conference of Mayors has adopted a policy position that parallels the Institute of Architect's stance. Part of the architectural and design community takes a consilient attitude with the idea of "biomimicry," an approach that copies natural solutions in designs to solve human problems. For example, engineers and architects collaborated on a building in Zimbabwe that minimized expenses for cooling by mimicking ventilation found in termite mounds. A product applying the principle of biomimicry is a cane for visually impaired people that uses ultrasonic sound to navigate in a fashion similar to that used by bats.[40]

Artists, writers, and musicians can also be involved in solving global environmental problems. From my perspective in the United States, the most prominent creative people are involved in popular television, movies, and music. Most of these media's productions are primarily preoccupied with people and very distant from nature. If the images we see on a daily basis do not allow for a place for nature in our lives, most people will think that there is no need to protect the environment that supports us. Social norms and expectations are determined in part by how the arts interpret and present the world.

A Proposal for Socioenvironmental Restoration

Paul R. Ehrlich and Donald Kennedy proposed that humans must assess their own behavior to appropriately respond to global environmental issues.[41] This assessment would allow the exchange of information about what societies and cultures around the world expect of global environmental protection and the ecological services that can be provided to them. The assessment would be only the start of a dialogue on human behavior that ultimately will be necessary to address global environmental problems.

The core of my proposal for socioenvironmental restoration involves forming a new idea of who owns and controls global ecological goods and services. As it stands in many societies, if people own land, they own the plants, animals, microorganisms, soils, and waters on the land; that is, they control the ecological goods and services on that land. Ownership of land and ecological goods and services could be separated. This separate ownership would be similar to mining rights in the United States; rights to mine under land are owned separately from the actual property. Similarly, ecological goods and services are a public trust and could be separated from ownership of land. If people had their own stake in the environment, then they would perceive those companies, governments, and individuals who used more than their share to be "cheaters" and would be more willing to control use of those resources.

The concept of private ownership of ecological goods and services has been explored more fully by Peter Barnes.[42] His first point is that, just like private citizens, corporations have legal rights in the United States and are entitled to the same protections under the Fourteenth Amendment. His second point is that because corporations control so much wealth, they have unprecedented political power. He proposes giving equal standing to ecological goods and services and having the public rights to these held in a trust. This legal standing will balance the unfair advantage that corporations have in society and politics. Barnes notes that Thomas Paine, the early American pamphleteer, presented a simple version of this proposal over 200 years ago. Recovering individual ownership of ecological goods and services is the aim of my proposal for socioeconomic restoration.

The following actions could provide a path through socioeconomic restoration to transition toward a stable existence of humans on our planet:

1. Reach wide consensus on goals to attain global environmental stability, including but not limited to stabilization of human population, resource use rates, and global climate change, while reducing human suffering, poverty, diseases, abuses of human rights, pollution, species extinctions, species introductions, and habitat destruction.

2. Where possible, establish a precedent for global individual ownership of ecological goods and services. That means each individual on Earth understands that he or she is entitled to a certain portion of ecological goods and services, and that no company or government has the right to take control of these from any person without permission or adequate compensation for their continuing value over time.

3. Establish conditions in which each individual can use resources only up to his or her own allotted ecological footprint. Accountability for resource use rates would force technological innovation to increase the standard of living without detriment to the environment.

4. Launch education initiatives based on principles of education that work. Education of young people is necessary to ensure that everyone has basic ecological knowledge of how our global ecosystems work. Specialists could be educated to work with other specialists (how to collaborate across disciplines). Consilience requires collaboration. A specialist is not likely to collaborate until he appreciates the possibilities and benefits of collaboration. Adult education should allow people to assess who is consuming ecological capital—for example, who is exploiting my allotted ecological footprint? Just as newspapers and the Internet publish daily details on stock prices, they could provide reports on ecological capital. If ecological goods and services are assigned real monetary value, news sources could provide education on how that value is distributed. Education will also create a social norm, so that

using more than your allotted resources becomes socially unacceptable, and regulation becomes more effective.

5. Create a political climate that can control those who use more than their share of global resources and provide incentives for those who use less than their share. The public will need to support environmental regulations in a general sense if they are to be effective. Regulations and incentives will need to be culturally acceptable, ecologically reasonable, and economically sound. Ecological scientists can assess how various human activities influence our environment, but decisions on how much damage is acceptable will be social, political, and cultural. Incentives and regulations that are not perceived as fair by most people are not as likely to be supported. An international approach that stresses equitability across cultures and societies will be required.

6. Use behavioral and economic principles that make regulations enforceable and not counterproductive. Policies that go against basic economic principles will be ineffective. We need to account for environmental externalities and move from using growth as an indicator of economic health to full environmental accounting for calculating gross domestic product in order to accomplish sustainable economies. Polling can be used to determine how large incentives for conservation should be and how stiff penalties for environmental crimes should be in order to be effective. Polling can account for fluctuation in values of ecosystem goods and services and make dynamic regulation possible. For example, as a species becomes rarer, it is assigned a higher value. Quantitative details of this approach are given in appendix 3.

7. Use the accumulated knowledge of humanity to preserve and restore ecological systems around the globe.

What Society Must Do to Limit Our Footprint

As discussed already, a number of basic steps are required to mitigate the human impacts on Earth's ability to support us and other species. The basic guiding principle is that human actions should allow for current and future sustenance of humans and other species that inhabit our planet. Human population size and rates of resource use need to be controlled. We must reduce the rates of release of greenhouse gases, fertilizers, toxins, CFCs, and other chemicals that destroy ozone. The protection or restoration of substantial areas containing representative biodiversity of all major habitats, in addition to areas vital to ecosystem function, is crucial. Controlling the rate of human-caused species invasions and halting species extinctions are imperative. Delaying action on any of these measures to protect the global environment will only exacerbate the problems. For example, delayed caps on greenhouse gases will make it more difficult to control the problem without more drastic action in the future.[43] Accomplishing these goals requires a substantial change in the aggregate behaviors of human society. Life on Earth is passing through a human-induced bottleneck, and we may not make it through to the other side.[44]

Success in making these changes will be more likely if the key aspects of human behavior described in chapters 5 and 6 are taken into account. Controlling fertility through birth control is more likely to be successful than attempting to limit frequency of copulation. Reducing release of pollutants will be more palatable if accomplished through increased efficiency of resource use and utilization of technological methods to reduce environmental impacts of resource use. Asking people to voluntarily lower their standard of living and comfort will not work. It will be easier to use education and change social norms that currently encourage display of wealth and resource use than to try to enforce laws that make excessive wealth and resource use illegal. Understanding how individuals in group situations assess costs and benefits of protecting the environment will increase the chances of successful control over environmental exploitation.

Solutions will require specialists from all areas of human endeavors. Scientists will provide technological solutions; economists will make available true cost-benefit analyses that account for all externalities, especially environmental impact costs; social scientists will establish how attempted behavioral changes fit within individual cultures and how individual choices will influence aggregate human behavior; religious leaders and ethicists will help provide the moral basis for preserving Earth; and educators will transmit ecological and environmental literacy to the public.

All these approaches require a major increase in funding available to the various specialties. This shift in resources toward environmental protection will need to at least rival some previous large commitments of money by society, such as those dedicated to space travel, major wars, and disease prevention. Relative to other areas, only a small fraction of societal efforts have gone toward environmental protection. If the U.S. government can pour massive amounts of funding into putting men on the moon and sequencing the human genome, it certainly can afford research that will lead to ensuring the future of our entire planet and human survival.

The Millennium Ecosystem Assessment has identified a number of policy recommendations that will improve or protect ecosystem services.[45] These include major investment in poverty reduction and public education and infrastructure, expanded markets for ecosystem services, elimination of trade barriers and subsidies that distort markets so they harm ecosystems, investment in technological innovation, and reorganization of institutions for adaptive governance. To accomplish these recommendations, governing institutions should (1) integrate ecosystem protection goals within a broader framework, including other sectors of government; (2) increase coordination of international multilateral agreements to consider environmental protection; and (3) increase transparency and accountability of government and private-sector institutions.

Underestimating the magnitude of the commitment required to reverse global environmental changes will doom efforts to failure. Nothing short of a complete change in the way global society views and reacts toward the environment that supports it, transcendence and enlightenment, will avert the major problems that will arise if we continue with the current attitude of "business as usual."

What Individuals Can Do to Limit Their Footprint

Humanity's approach to our environment will not change unless individuals work on altering the current system. Political leaders maintain power by pleasing individuals. If politicians perceive that the public is interested in protecting the global environment to the point where they will support only leaders who protect our Earth, leaders will promote environmental protection as the only means to retaining their power. If working for a sustainable Earth will not earn votes, efforts of the politicians will go toward pleasing corporate donors and any others with a financial stake in continued exploitation of Earth's resources.

The basics of human nature related to environmental issues make global environmental damage an extremely difficult problem to solve, and any solution will require the individual efforts of billions of people. Given the magnitude of global environmental change, and the difficulty of effecting such a change, dedication will be necessary to control the momentum imparted by the environmental catapult. A first step would be to minimize individual resource use rates.

Relatively easy alterations of personal habits can substantially decrease the impact that each individual has on the environment. The Global Footprint Network has developed a calculator that can be used to assess the area of the Earth's surface it takes to support each person: the ecological footprint. For example, "Bob" in Chicago, who eats meat daily, as well as mostly processed and packaged foods; lives in a 2000-square-foot (184-m^2) house with three other people; drives 200 to 300 miles (322 to 483 km) per week in a car that gets 15 to 25 miles per gallon (6.3 to 10.6 km per L); rarely bicycles, walks, or uses public transportation; and uses 10 hours per year of air travel, has a footprint of 35 acres (14 ha). Bob can cut his footprint in half by eating meat only occasionally, consuming less packaged and processed food, recycling, living in a house three-quarters the size, using public transportation about one-half the time, biking and walking occasionally, driving a car with double the gas mileage, and carpooling. Many of those living in the United States can make lifestyle changes of this magnitude without endangering their health or happiness.

Dedication to minimizing personal resource use rates can also be extended by limiting individual reproductive output; writing letters or sending printed information (even a copy of this book) to public officials; educating friends, family, and others about potential global environmental problems; writing letters to news media; and donating money to lobbying groups and to organizations that promote the control of population growth and resource use rates (appendix 4).

Tell people you know about the threats described in this book; let them know how the trends could compromise the health and well-being of their offspring. You can do your part to educate society, and education is the key to successful political and economic action. Individuals educating their peers can be more effective than education from other sources. Modeling, or seeing others' behavior, increases education effects by enforcing a social norm. The behavior you exhibit will influence others.

Get involved in the local political process. Global environmental impacts are the aggregate of thousands of local political decisions. In the United States, states are taking the lead in controlling greenhouse gas emissions, even if the national government is exhibiting no real leadership on this issue.

For young readers or those contemplating job changes, directing career choices toward solving the problems mentioned in this book could help produce the technological, religious, or social advances necessary to transform us into a global, environmentally sustainable society. Many routes could be followed to accomplish this because education, political action, altered economic approaches, changed religious attitudes, and scientific advances all need to be implemented if the global environmental problems we are faced with will ever be solved. Even if many individuals make all these changes, it might not have an effect. But an attempt to leave the world's environment better is positive action and far superior to ignoring or exacerbating a global crisis.

The Dalai Lama makes a persuasive case for the need to control the size of humanity's footprint. He argues for the interconnectedness of humanity and science and its conceptual frameworks, offering, in my opinion, the kind of inspiration that humanity needs:

> The earth is our only home. As far as current scientific knowledge is concerned, this may be the only planet that can support life. One of the most powerful visions I have experienced was the first photograph of the earth from outer space. The image of a blue planet floating in deep space, glowing like the full moon on a clear night, brought home powerfully to me the recognition that we are indeed all members of a single family sharing one little house. I was flooded with the feeling of how ridiculous are the various disagreements and squabbles within the human family. I saw how futile it is to cling tenaciously to the differences that divide us. From this perspective one feels the fragility, the vulnerability of our planet and its limited occupation of a small orbit sandwiched between Venus and Mars in the vast infinity of space. If we do not look after this home, what else are we charged to do on this earth?[46]

Appendix 1

Data Sources Used to Make Graphs

TABLE 1 Population Growth and Resource Use Rates on Earth

	A	B	C	D	E	F	G
Year	Population (millions)	Average world gross domestic product (1990 U.S. $ per person)	Total gross domestic product (1990 U.S. $trillion)	Energy use (million metric tons of oil equivalents per year)	Energy use per person (metric tons of oil per person per year)	Global water use (km³ per year)	Per capita water use (m³ per person per year)
1800	**917**	459	0.42	319	0.35	186	203
				(400)		**(243)**	
1820	991	577	0.57	502	0.51	253	255
		(651)					
1830	1066	648	0.69	629	0.59	295	277
1840	1140	727	0.83	788	0.69	344	302
1850	**1214**	815	0.99	988	0.81	401	330
1860	1300	915	1.19	1239	0.95	468	360
1870	1386	1027	1.42	1553	1.12	545	393
1880	1472	1152	1.70	1946	1.32	636	432
1890	1558	1293	2.01	2440	1.57	742	476
1900	**1644**	1450	2.38	3058	1.86	865	526
		(1263)		**(1900)**		**(580)**	
1910	1730	1627	2.82	3833	2.22	1009	583
1920	**1813**	1826	3.31	4805	2.65	1177	649

(*continued*)

TABLE 1 (*continued*)

	A	B	C	D	E	F	G
Year	Population (millions)	Average world gross domestic product (1990 U.S. $ per person)	Total gross domestic product (1990 U.S. $trillion)	Energy use (million metric tons of oil equivalents per year)	Energy use per person (metric tons of oil per person per year)	Global water use (km³ per year)	Per capita water use (m³ per person per year)
1925	1896	1934	3.67	5380	2.84	1271	670
1930	**1987**	2049	4.07	6023	3.03	1372	690
1935	2073	2170	4.50	6743	3.25	1482	715
1940	**2213**	2299	5.09	7550	3.41	1601	723
1945	2317	2435	5.64	8453	3.65	1729	746
1950	**2496**	2580	6.44	9464	3.79	1867	748
		(2138)				**(1360)**	
1955	**2752**	2733	7.52	10,595	3.85	2016	733
1960	**3019**	2894	8.74	11,862	3.93	2177	721
1965	**3336**	3066	10.23	13,281	3.98	2351	705
1970	**3676**	3248	11.94	14,869	4.04	2539	691
						(2590)	
1975	**3995**	3440	13.74	16,647	4.17	2742	686
1980	**4438**	3644	16.17	18,638	4.20	2962	667
1985	**4851**	3860	18.73	20,867	4.30	3198	659
1990	**5313**	4089	21.72	23,362	4.40	3454	650
		(5145)		**(30,000)**		**(4130)**	
1995	**5691**	4331	24.65	26,156	4.60	3730	655
2000	**6080**	4588	27.90	29,284	4.82	4028	663
						(5190)	
2005	**6482**	4845	31.15	32,412	5.04	4400	679

NOTE: Bold values are the published values used for extrapolations.

SOURCES: (*A*) World population from J. E. Cohen, *How Many People Can the Earth Support?* (New York: Norton, 1995). Extrapolations made by eye to fill in numbers between published values. Recent values from the U.S. Census Bureau. (*B*) Bold per capita domestic product values from J. R. McNeill, *Something New Under the Sun: An Environmental History of the Twentieth-Century World* (New York: Norton, 2000). Extrapolation made with linear regression of natural log transformed data, $r^2 = .92$. (*C*) Column A multiplied by column B. (*D*) Bold per capita energy values from McNeill, *Something New Under the Sun*. Extrapolation made with linear regression of natural log transformed data, $r^2 = .93$. (*E*) Column D divided by column A. (*F*) Bold water use values from McNeill, *Something New Under the Sun*. Extrapolation made with linear regression of natural log transformed data, $r^2 = .92$. (*G*) Column F divided by column A.

TABLE 2 Carbon Dioxide Data

Year	Carbon dioxide (parts per million)	Year	Carbon dioxide (parts per million)
1800	280	1950	311
1820	284	1960	317
1850	288	1965	320
1890	297	1970	326
1900		1975	331
1910		1980	339
1920		1985	346
1930		1990	354
1940		1995	361
		2000	370
		2005	378

SOURCES: Data up to 1950, from Siple Station Ice Core (A. Neftel et al., Physics Institute, University of Bern, Switzerland); from 1960 on, from Mauna Loa Observatory, Hawaii (C. D. Keeling and T. P. Worf, Scripps Institute of Oceanography, University of California, La Jolla, California). Plotted graph (figure 3.1) is the linear extrapolation between these values.

TABLE 3 Forest, Cropland, and Human Population

	A	B	C	D	E
Year	Population (million)	Cropland (million km²)	Forest (million km²)	Cropland per person (km²)	Urban population (%)
1700	600	4	54	0.0068	
1710	606	4	54	0.0072	
1720	612	5	54	0.0075	
1730	617	5	53	0.0079	
1740	623	5	53	0.0083	
1750	629	5	53	0.0087	
1760	687	6	53	0.0083	
1770	744	6	53	0.0081	
1780	802	6	52	0.0078	
1790	859	7	52	0.0076	
1800	917	7	52	0.0075	
1810	976	7	52	0.0073	
1820	1036	7	52	0.0072	
1830	1095	8	51	0.0070	
1840	1155	8	51	0.0069	
1850	1214	8	51	0.0068	
1860	1300	9	51	0.0069	
1870	1386	9	50	0.0068	
1880	1472	10	50	0.0069	
1890	1558	11	50	0.0070	14
1900	1644	12	49	0.0070	
1910	1729	12	49	0.0072	18

(continued)

TABLE 3 (*continued*)

Year	A Population (million)	B Cropland (million km^2)	C Forest (million km^2)	D Cropland per person (km^2)	E Urban population (%)
1920	1813	13	49	0.0072	
1930	1987	14	48	0.0072	23
1940	2213	15	48	0.0068	
1950	2496	15	48	0.0062	29
1960	3019	17	47	0.0056	
1970	3676	17	47	0.0047	37
1980	4438	18	47	0.0041	
1990	5313	18	47	0.0034	
1992	5451	18	47	0.0033	43
1993	5533	18	46	0.0033	
1994	5613	18	46	0.0032	
1995	5694	18	46	0.0032	45
1996	5773	18	46	0.0032	
1997	5852	18	46	0.0031	
1998	5930	18	46	0.0031	
1999	6006	18	46	0.0030	
2000	6082	18	46	0.0030	47
2001	6156	18	46	0.0030	
2002	6230	18	46	0.0029	
2003	6303	18	46	0.0029	
2004					
2005					49

SOURCES: (A) World population from J. E. Cohen, *How Many People Can the Earth Support?* (New York: Norton, 1995). Extrapolations made by eye to fill in numbers between published values. Recent values from the U.S. Census Bureau. (B) Values from N. Ramankutty and J. A. Foley, "Estimating Historical Changes in Global Land Cover: Croplands from 1700 to 1992," *Global Biogeochemical Cycles* 13 (1999): 997–1027, and United Nations Food and Agriculture Organization. (C) Values from Ramankutty and Foley, "Estimating Historical Changes in Global Land Cover," and United Nations Environmental Program. (D) Column C divided by column A, table 1. (E) Data from United Nations and various other sources.

TABLE 4 Fertilizer, Pesticide, and Ozone Data

	A	B	C	D	E	F	G
Year	Ozone minimum (Dobson units)	Maximum area of ozone hole (million km²)	Ultraviolet index	Total fertilizer use (million metric tons)	Nitrogen (million metric tons)	Phosphorus (million metric tons)	Annual pesticide exports (million $)
1961				31	12	11	267
1962				34	13	12	304
1963				38	15	13	355
1964				42	16	14	398
1965				47	19	16	399
1966				52	22	17	473
1967				56	24	18	519
1968				60	26	19	560
1969				63	28	20	655
1970				69	32	21	738
1971				73	34	22	769
1972				79	36	24	880
1973	•			85	39	26	1273
1974				82	38	24	1774
1975				91	44	26	2303
1976				95	45	27	2225
1977				101	49	29	2504
1978			9.9	109	54	30	3246
1979	209	0.48	9.8	112	57	31	3806
1980	205	1.56	10.5	117	61	32	4467
1981	205	1.43	10.4	115	60	31	4271
1982	189	7.82	10.0	115	61	31	4221
1983	169	10.2	10.4	126	68	33	4506
1984	154	12.3	10.6	131	71	34	5130
1985	146	1704	11.2	129	70	33	5200
1986	159	13.02	10.8	132	71	35	5831
1987	120	21.65	10.8	140	76	37	6671
1988	158	12.35	10.9	146	80	38	7303
1989	124	20.95	10.9	143	79	37	7305
1990	128	20.17	11.2	138	77	36	8276
1991	117	22.05	10.8	134	75	35	7935
1992	124	24.47	11.3	126	74	31	8088
1993	126	25.14	11.5	121	73	29	8022
1994	88	24.63	11.3	122	73	30	8625
1995	98	23.05	11.2	129	78	31	10,276
1996	111	26.12	11.5	134	82	31	11,193
1997	104	24.43	11.8	137	81	33	11,085
1998	90	27.35	12.1	137	82	33	11,604
1999	95			147	89	33	11,184
2000	85	26.1		144	86	32	11,149
2001	88	25.4		147	87	34	10,387

(*continued*)

TABLE 4 (*continued*)

	A	B	C	D	E	F	G
Year	Ozone minimum (Dobson units)	Maximum area of ozone hole (million km²)	Ultraviolet index	Total fertilizer use (million metric tons)	Nitrogen (million metric tons)	Phosphorus (million metric tons)	Annual pesticide exports (million $)
2002	132	19.7		148	88	34	10,912
2003	110	28					12,574
2004	135	19					15,589
2005		25					

SOURCES: (*A* and *B*) Data from National Aeronautics and Space Administration (http://toms.gsfc.nasa .gov/multi/oz_hole_area.jpg) and European Global Ozone Monitoring experiments. (*C*) Data from Lauder, New Zealand; noontime summer maximum (R. McKenzie, G. Bodeker, and B. Connor, "Increased UV Radiation in New Zealand: A Cautionary Tale," *NIWA Water and Atmosphere* 7 [1999]: 8–9). (*D–G*) Data from United Nations Food and Agriculture Organization.

Appendix 2

Reading the Graphs in This Book

Many people are very comfortable reading graphs and effectively extracting information from them. However, in my decades of teaching college students, I have encountered numerous well-educated adults who have not yet developed this skill. The ability to read graphs well is critical in appraising information, as well as determining when a source is using complex graphics to obfuscate a point. I hope this appendix helps those who would like some more background in reading simple graphs.

Graphs are used to quickly convey the degree of relationship between values, as well as the variation in that relationship. Data can be organized according to categories or continuously. *Categories* might include types of things (for example, habitats or companies). *Continuous* organization shows values in relationship to each other (that is, on a numerical scale). Reading most graphs requires attention to two dimensions. Data (more than one datum) organized according to continuous scales in two dimensions are called *plots*. The graphs presented in this book mostly represent data along two continuous scales.

When continuous data are presented across two dimensions, there are two axes (ranges of values). The line that is horizontal across the bottom is called the x-axis; the values it represents increase from left to right. The vertical line to the left is the y-axis; its values typically increase from

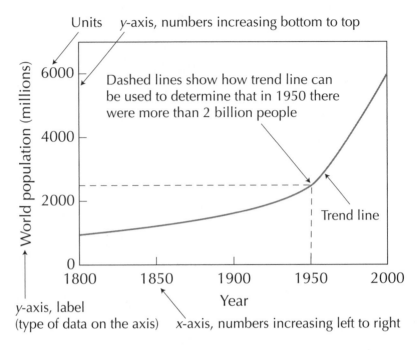

Units — y-axis, numbers increasing bottom to top

World population (millions)

6000 — Dashed lines show how trend line can be used to determine that in 1950 there were more than 2 billion people

4000

2000

Trend line

0

1800   1850   1900   1950   2000

Year

y-axis, label
(type of data on the axis)   x-axis, numbers increasing left to right

FIGURE A.1 An example graph with annotations of some important features as described in the appendix. The graph illustrates how you can find a y value that corresponds with any individual x value. The graph is figure 2.1.

bottom to top (figure A.1). Usually, the values on continuous scales are numbered sequentially, with even divisions between the labels (for example, 2, 4, 6, 8). These values are labeled along the axis so the relative position in the plot can be assigned a value on the scale. Each axis should be labeled with text. This label usually has two parts. The first part is what is being plotted (that is, the type of data). The second part of the label (usually in parentheses) is the unit that is being plotted. Units can be very important indicators of scale; a car traveling 1 mile (1.6 km) per hour is moving much less quickly than one moving 1 mile per second.

Nonlinear scales can be used at times. For example, a logarithmic scale is useful for plotting data that range widely. A common example of a logarithmic scale is the Richter scale for measuring earthquakes. Each increase of one unit (for example, from 5 to 6) indicates a tenfold increase in severity of earthquake. Sometimes, the scales on the axes range from zero to a maximum; other times, they are a range of numbers. It is important to pay attention to this. For example, a number that varies from 100 to 102 (only 2 percent) over five years can be made to appear to vary widely if a scale from 100 to 102 is used on the y-axis. If the same plot is presented with the values plotted with a range of 0 to 102, there will be very little apparent variation.

Most of the plots in this book are scatter plots or line plots. A *scatter plot* consists of a series of points that are arranged relative to their values in two dimensions. A *line plot* has a line that indicates a trend in the data. In a scatter plot, each point on the graph is assigned an x and a y value. Thus, when reading a graph, it is possible to use the x- and y-axes to assign two values to the point. Likewise, in a line plot, each location along the line has a corresponding x and y value.

The values plotted on the x-axis generally are considered the *independent variable*. That is, they do not vary as a function of the values on the y-axis (the *dependent variable*). The graph of world population is an example of this (see figure 3.1). Time, in years, has nothing to do with the number of humans on earth. A year goes by the same, regardless of how many people exist. In contrast, the amount of time does constrain how large a population can become because time is required for a population to grow.

The values plotted on the y-axis usually are considered dependent on the values on the x-axis. This implies that y values vary as a function of the x-axis. This x–y plotting is a convenient way to present data. The plotting allows for analysis of the relationship between two variables. For example, there are numerous statistical techniques that can be used to establish how strongly two variables correlate with each other.

A word of caution is necessary. Correlation is not causation. An example of this is the number of churches in each town plotted against the number of establishments where alcohol is served. There is a definite trend: New York City has lots of churches and bars; Little Rock, Arkansas, has fewer; and a town with 1000 people usually has no more than a few of either. It is difficult to prove that bars cause churches, or vice versa. Regardless of how strong the correlation, it is not causation. Nevertheless, correlation is a powerful tool. Smoking rates and lung cancer rates are closely correlated. The tobacco companies argued for years that the correlation does not prove causation. That correlation does not *prove* causation is true—a specific case of lung cancer cannot be absolutely tied to smoking tobacco. But many other factors have been considered and ruled out (for example, age, sex, geographic location). After other potential causes are considered, correlation eventually makes a strong case for causation.

Other types of graphs also are used on occasion. For example, bar graphs generally plot a continuous variable on the y-axis against categories on the x-axis. Pie charts use the size of the pie slice to represent the continuous y-axis variable. Three-dimensional graphs can be used to represent more complex data (although three-dimensional bar and pie charts are often really two-dimensional data fancied up to three dimensions and rendered less readable).

When viewing graphs, the interpreter must be careful and aware. If the art of viewing graphs is well developed, a vital tool of data analysis has been mastered. A graph, like a picture, can be worth a thousand words.

Appendix 3
Putting Global Environmental Impact into a Quantitative Framework That Links Impact and Behavior

The impact of people on environment is a function of the number of people and how much impact each person has. This relationship was initially proposed in the 1970s as $I=PAT$, where I is impact, P is population, A is affluence, and T is technology. More recently, the relationship was expanded to the ImPACT equation, $I=PACT$, where C is the intensity of use per good per affluence.[*] For example, the impact of air pollution from driving cars is I (tons of pollutant)$=P$ (number of people driving)$\times A$ (miles driven per person)$\times C$ (gallons burned per mile)$\times T$ (tons of pollutant per gallon). In this example, I can be regulated by decreasing the number of people driving (car pooling), decreasing the distance each person drives, increasing gas mileage, or decreasing the amount of pollutant released per unit of gas burned.

Now let's say that we try to change the tons of pollutant generated, focusing on T, tons of pollutant per gallon. We can enact laws that require cleaner fuel to be burned, but this fuel is more expensive. So there are two options for each person: either comply with the law or break the law and use cheaper dirty fuel. This is where game theory comes in. A certain

[*]P. E. Waggoner and J. H. Ausubel, "A Framework for Sustainability Science: A Renovated IPAT Identity," *Proceedings of the National Academies of Sciences* 99 (2002): 7860–7865.

number of people (P_c) will cooperate and use the clean fuel, and the rest will be selfish (P_s) and use dirty fuel. The total number of people, $P = P_s + P_c$, is divided into two groups, and we will need two full sets of calculations for tons of pollutant: one that quantifies the effect of selfish people and the other for cooperators. The idea is to make cooperation the predominant strategy. Perhaps the easiest way to do this is to subsidize the expensive clean fuel so it costs less than dirty fuel. If this solution is too expensive, then another alternative is to punish those who are caught using dirty fuel. The greater the penalty, the fewer people will cheat and use dirty fuel. If the penalty is less than the difference in price between clean and dirty fuel, then most will use the dirty fuel. The proportion of people who think the risk of getting caught using dirty fuel is worth the gamble because of the money they save becomes a function of the relative cost of cooperating and the potential costs of getting caught. If we define the proportion of the population that thinks there is a net benefit (benefits exceed costs) to co-operating as B_c, we can calculate the absolute number of cooperators, $P_c = P \times B_c$, as well as P_s.

It is necessary to define the proportion of those who perceive benefits of cooperating, B_c, under a variety of conditions. This could be accomplished by surveys of responses or by observations of actual behavior under different imposed costs and benefits. For example, different penalties for cheating could be enacted in different areas. Then the perceived benefit of cheating relative to the potential cost of the penalty could be based on the observed proportion of cheaters in each area (in the case of air pollution and dirty and clean fuel, the amount of air pollution that cannot be accounted for, assuming that all are cooperating and using clean fuel). Society will bear a cost of influencing B_c because assessing both penalties for cheating and benefits for cooperating incur a burden on society. A cost-benefit calculation is needed that balances costs of encouraging cooperation against societal benefits that are accrued from cooperation. Stated another way, governments need to determine what costs they are willing to burden the public with against the benefits of altering public behavior. Risk analysis comes into play here. If there is even a small risk of catastrophe, then greater costs to society for ensuring cooperation are acceptable to more people.

The ImPACT variables allow individual behavior to be altered to have an influence. For example, technology has more of a direct influence on C and T, whereas behavioral modifications directly affect P and A. In the tons of pollutant example, P can be influenced by decreasing the number of people who drive (by mass transit or carpooling), and A can be influenced by altering the number (or length) of trips each person takes. Establishment of social norms—for example, increasing the cost of gasoline or automobiles—can alter these factors. And regulation can alter the factors—for example, issuing a fixed number of drivers' licenses or allowing a fixed number of vehicles to be sold. These are solutions that rely on social, and not technological, factors. In contrast, C and T can be influenced by increasing fleet mileage or producing cleaner fuel. As discussed, getting people to use the improved technology may require behavioral modification as well.

Sometimes, regulation may depend on modifying the behavior of only a small number of people, if those people have a disproportionate effect on per capita rates of pollution. Put more specifically, if the mean per capita rate of pollution is far greater than the median, then a few individuals responsible for the greatest amount of pollutant release have a disproportionate effect on the total rate of pollution. For example, although automobile emissions testing has been required in some areas to control air pollution, researchers have found that only a few automobiles were causing the bulk of the pollution. This was discovered with ingenuous technology, using lasers and sensors that could detect pollutants issuing from cars in traffic. In this case, it is more efficient to focus regulation on the few automobile owners with cars that are polluting heavily and avoid expensive and time-consuming emission testing for most cars. Mean per capita rates of pollution are less meaningful in this case. Likewise for greenhouse gas emissions, focusing on people in countries or specific cities with the greatest emission rates per capita is an efficient allocation of effort.

The relatively simple equation becomes particularly complex when we consider global environmental problems. A full set of equations may be necessary for each problem. For global carbon emissions, we may use sets of equations for automobile use, industrial burning, and deforestation. In some countries, industry may be efficient; in others, it may not. Each country, or maybe even each region of each country, would require a separate set of equations. Each of the terms of the equation can vary, depending on a number of factors, and some factors may be interrelated. In the preceding example, the proportion of people cheating and using cheap dirty gas may decrease if fuel-efficient vehicles are available at modest cost, because the total fuel costs decrease per mile driven. Behaviors will also vary across and even within countries. As complex as these calculations could become, they are not so complex that teams of analysts could not accomplish them.

The mitigation of impact requires scientists to modify technology, policy makers to implement solutions, and social scientists to guide the policy into effective paths. Ultimately, decisions will also be driven by moral issues: for example, is it ok to use our children's' resources for today's problems? The framework presented here gives hope that an analytical approach to environmental problems may help solve issues that revolve around both technological and behavioral factors.

Appendix 4

Some Organizations Involved
with Global Environmental Issues

International Planned Parenthood Federation
Regent's College
Inner Circle, Regent's Park
London NW1 4NS
United Kingdom
Phone: +44 (0) 20-7487-7900
Fax: +44 (0) 20-7487-7950
Web site: http://www.ippf.org

The International Planned Parenthood Federation promotes and defends the right for women, men, and young people to freely decide the number and spacing of their children, as well as the right to the highest possible level of sexual and reproductive health. Through information, advocacy, and services, the federation works cooperatively with interested governmental and nongovernmental parties to advance the idea of family planning, maternal and child health, and, notably, the elimination of unsafe abortion. It is committed to obtaining equal rights for women and empowering women to obtain full participation in, and benefit from, social and economic development. The federation relies on committed, competent, skilled volunteers and staff of member associations to provide leadership in the execution of their mandates.

League of Conservation Voters
1920 L Street, NW
Suite 800
Washington, D.C. 20036
Phone: (202) 785-8683
Fax: (202) 835-0491
Web site: http://www.lcv.org

The League of Conservation Voters, an American nonpartisan political organization, is dedicated to electing members to Congress who are pro-environment. The group works through political action to protect the environment by holding Congress and the administration accountable for their decisions. It supports programs including the National Environmental Scorecard, CongressWatch, and Eye on the Administration.

*Native*Energy
823 Ferry Road
P.O. Box 539
Charlotte, Vt. 05445
Phone: 800.924.6826
E-mail: info@nativeenergy.com
Web site: http://www.nativeenergy.com

*Native*Energy, a privately held Native American energy company, provides information and equipment to encourage the use of wind power and dairy farm methane energy as primary energy sources. In August 2005, the Intertribal Council on Utility Policy (COUP) acquired a majority ownership interest in *Native*Energy on behalf of its member tribes. The organization states: "Even if you drive a hybrid car, use compact fluorescent lighting, and take every energy-saving step you can, you are still going to use some energy and create CO_2 pollution. There's only one way to do something about the CO_2 pollution you can't avoid creating—help reduce CO_2 pollution another way."

The Nature Conservancy
4245 North Fairfax Drive
Suite 100
Arlington, Va. 22203-1606
Phone: (800) 628-6860
E-mail: comment@tnc.org
Web site: http://www.nature.org

A leading conservation organization, the Nature Conservancy works around the world to protect ecologically important lands and waters for nature and people. Founded in 1951, the organization has projects in all 50 states and more than 30 countries. Its mission is "to preserve the plants, animals, and natural communities that represent the diversity of life on Earth by protecting the lands and waters they need to survive."

Population Connection
2120 L Street, NW
Suite 500
Washington, D.C. 20037
Phone: (202) 332-2200; (800) 767-1956
Fax: (202) 332-2302
E-mail: info@populationconnection.org
Web site: http://www.populationconnection.org

Population Connection, a national nonprofit organization, works to slow population growth and "achieve a sustainable balance between the Earth's people and its resources." The goal is to protect the environment and thereby ensure a high quality of life for present and future generations. The organization's education and advocacy programs are designed to "influence public policies, attitudes, and behavior on national and global population issues and related concerns."

Rocky Mountain Institute
1739 Snowmass Creek Road
Snowmass, Colo. 81654-9199
Phone: (970) 927-3851
E-mail: dalelevy@rmi.org
Web site: http://www.rmi.org

The goal of Rocky Mountain Institute, an entrepreneurial, nonprofit organization, is the "efficient and restorative use of resources to create a more secure, prosperous, and life-sustaining world." It "show[s] businesses, communities, individuals, and governments how to create more wealth and employment, protect and enhance natural and human capital, increase profit and competitive advantage, and enjoy many other benefits— largely by doing what they do far more efficiently." The institute's work is "independent, nonadversarial, and transideological, with a strong emphasis on market-based solutions."

Sierra Club
85 Second Street
Second Floor
San Francisco, Calif. 94105-3441
Phone: (415) 977-5500
Fax: (415) 977-5799
Web site: http://www.sierraclub.org

The motto of the Sierra Club is "Explore, enjoy, and protect the planet." It is estimated that there are more than 1.3 million members, and the organization is "America's oldest, largest, and most influential grassroots environmental organization." It encourages the responsible use of the Earth's ecosystems and resources, using education and lawful means to enlist people to "protect and restore the quality of the natural and human environment."

Union of Concerned Scientists
2 Brattle Square
Cambridge, Mass. 02238-9105
Phone: (617) 547-5552
Fax: (617) 864-9405
Web site: http://www.ucsaction.org

The Union of Concerned Scientists seeks to ensure that all people have clean air, energy and transportation, as well as food that is produced in a safe and sustainable manner. Begun in 1969, the union is now an alliance of more than 200,000 citizens and scientists, using independent scientific analyses to "secure changes in government policy, corporate practices, and consumer choices that will protect and improve the health of our environment globally, nationally, and in communities throughout the United States." Its achievements over the decades show that thoughtful action based on the best available science can help safeguard our future and the future of our planet.

U.S. Committee for United Nations Population Fund
220 East Forty-second Street
New York, N.Y. 10017
Phone: (212) 297-5210
Fax: (212) 297-5209
Web site: http://www.unfpa.org

The United Nations Population Fund promotes the right of every person to "enjoy a life of health and equal opportunity." Using population data for policies and programs, the organization helps developing countries find solutions "to reduce poverty and to ensure that every pregnancy is wanted, every birth is safe, every young person is free of HIV/AIDS, and every girl and woman is treated with dignity and respect." After beginning operations in 1969, it has become the largest international source of population assistance.

The World Conservation Union
Rue Mauverney, 28
1196 Gland
Switzerland
E-mail: membership@iucn.org
Web site: http://www.iucn.org

The mission of the World Conservation Union is "to influence, encourage, and assist societies throughout the world to conserve the integrity and diversity of nature and to ensure that any use of natural resources is equitable and ecologically sustainable." A multicultural, multilingual organization, the union is an important conservation network worldwide, with 1100 staff members located in 40 countries. Founded in 1948, the organization's earlier name was the International Union for the Conservation of Nature and Natural Resources.

World Wildlife Fund
1250 Twenty-fourth Street, NW
P.O. Box 97180
Washington, D.C. 20090-7180
Phone: (800) CALL-WWF
Fax: (202) 293-9211
Web site: http://www.wwf.org

Known worldwide by its panda logo, the World Wildlife Fund is dedicated to protecting the world's wildlife and wildlands. The largest privately supported international conservation organization in the world, it has more than 1 million members in the United States alone. Since its inception in 1961, the organization has invested in over 13,100 projects in 157 countries. WWF directs its conservation efforts toward three global goals: protecting endangered spaces, saving endangered species, and addressing global threats. From working to save the giant panda, tiger, and rhino to helping establish and manage parks and reserves worldwide, WWF has been a conservation leader for 40 years.

Worldwatch Institute
1776 Massachusetts Avenue, NW
Washington, D.C. 20036-1904
Phone: (202) 452-1999
Fax: (202) 296-7365
E-mail: worldwatch@worldwatch.org
Web site: http://www.worldwatch.org

The Worldwatch Institute is dedicated to fostering the evolution of an environmentally sustainable society—one in which human needs are met in ways that do not threaten the health of the natural environment or the prospects for future generations. Using interdisciplinary, nonpartisan research and believing that information is a powerful tool for social change—that human behavior shifts in response to either new information or new experiences—the institute seeks to disseminate information to bring about the changes needed to build an environmentally sustainable economy. Given the Earth's unified ecosystem and an increasingly integrated global economy, only a global approach to issues such as climate change, depletion of the stratospheric ozone layer, the loss of biological diversity, degradation of oceans, and population growth can be effective.

Notes

Preface

1. J. E. Cohen, *How Many People Can the Earth Support?* (New York: Norton, 1995).

1. Collision Course

1. R. C. Thompson, Y. Olsen, R. P. Mitchell, A. Davis, S. J. Rowland, et al., "Lost at Sea: Where Is All the Plastic?" *Science* 304 (2004): 838.

2. M. Wackernagel, N. B. Schulz, D. Deumling, A. Callejas Linares, M. Jenkins, et al., "Tracking the Ecological Overshoot of the Human Economy," *Proceedings of the National Academy of Sciences* 14 (2002): 9266–9271.

3. B. McKibben, *The End of Nature* (New York: Anchor, 1999).

4. P. M. Vitousek, P. R. Ehrlich, A. H. Ehrlich, and P. A. Matson, "Human Appropriation of the Products of Photosynthesis," *BioScience* 36 (1986): 368–373; P. M. Vitousek, H. A. Mooney, J. Lubchenco, and J. M. Melillo, "Human Domination of Earth's Ecosystems," *Science* 277 (1997): 494–499; P. M. Vitousek, H. J. Aber, R. W. Howarth, G. E. Likens, P. A. Matson, et al., "Human Alteration of the Global Nitrogen Cycle: Causes

and Consequences," *Issues in Ecology, Ecological Society of America* 1 (1997): 1–15.

5. Millennium Ecosystem Assessment, *Ecosystems and Human Well-Being: Synthesis* (Washington, D.C.: Island Press, 2005).

6. E. Callenbach, *Ecology: A Pocket Guide* (Berkeley: University of California Press, 1988).

7. M. Hertsgaaard, *Earth Odyssey: Around the World in Search of Our Environmental Future* (New York: Broadway Books, 1998).

8. Gallup International, "Millennium Survey" (available at: http://www.gallup-international.com/ContentFiles/millennium11.asp).

9. F.J. Mata, L. Onisto, and J. Vallentyne, *Consumption: The Other Side of Population for Development* (Cairo: International Conference on Population and Development, 1994).

10. P.R. Ehrlich, *The Population Bomb* (New York: Ballantine, 1968).

11. P.H. Raven, "Science, Sustainability, and the Human Prospect," *Science* 297 (2002): 954–958, at 958.

2. The Insidious Explosion

1. U.S. Census Bureau (available at: http://www.census.gov).

2. T.R. Malthus, *An Essay on the Principle of Population* (1798; repr., London: Macmillan, 1926).

3. W.R. Catton Jr., *Overshoot: The Ecological Basis of Revolutionary Change* (Urbana: University of Illinois Press, 1980).

4. J.E. Cohen, *How Many People Can the Earth Support?* (New York: Norton, 1995).

5. J. Diamond, *Collapse: How Societies Choose to Fail or Succeed* (New York: Viking, 2005).

6. Cohen, *How Many People Can the Earth Support?*

7. United Nations, Department of Economic and Social Affairs, Population Division, "World Population Prospects: The 2000 Revision," 2001, and "World Urbanization Prospects: The 2001 Revision," 2002 (available at: http://esa.un.org/unpp).

8. United Nations, Department of International Economic and Social Affairs, *Long-Range World Population Projections: Two Centuries of Population Growth, 1950–2150* (New York: United Nations, 1992).

9. Ibid.

10. United Nations, Department of Economic and Social Affairs, Population Division, "World Population Prospects: The 2006 Revision"

(available at: http://www.un.org/esa/population/publications/wpp2006/wpp2006.htm).

11. J. E. Cohen, "Population Growth and the Earth's Human Carrying Capacity," *Science* 269 (1995): 341–346.

12. J. Holdren, "Population and the Energy Problem," *Population and Environment* 12 (1991): 231–255.

13. F. Pearce, *When the Rivers Run Dry: Water—The Defining Crisis of the Twenty-first Century* (Boston: Beacon, 2006).

14. S. L. Postel, G. C. Daily, and P. R. Ehrlich, "Human Appropriation of Renewable Fresh Water," *Science* 27 (1996): 785–788.

15. R. Costanza, R. d'Arge, R. de Groot, S. Farberk, M. Grasso, et al. "The Value of the World's Ecosystem Services and Natural Capital," *Nature* 387 (1997): 253–260.

16. N. E. Bockstael, A. M. Freeman, R. J. Kopp, P. R. Portnoy, and V. K. Smith, "On Measuring Economic Values for Nature," *Environmental Science and Technology* 34 (2000): 1384–1389.

17. T. Prugh, R. Costanza, and H. Daly, *The Local Politics of Global Sustainability* (Washington, D.C.: Island Press, 2000).

18. R. Stone and H. Jia, "Going Against the Flow," *Science* 313 (2006): 1034–1037.

19. W. K. Dodds, *Freshwater Ecology: Concepts and Environmental Applications* (San Diego: Academic Press, 2002).

20. N. Johnson, C. Revenga, and J. Echeverria, "Managing Water for People and Nature," *Science* 292 (2001): 1071–1072.

21. D. Pimentel, B. Berger, D. Filiberto, M. Newton, B. Wolfe, et al., "Water Resources: Agricultural and Environmental Issues," *BioScience* 54 (2004): 909.

22. F. Tao, M. Yokozawa, Y. Hayashi, and E. Lin, "Terrestrial Water Cycle and the Impact of Climate Change," *Ambio* 32 (2003): 295.

23. N. L. Poff, M. M. Brinson, and J. W. Day Jr., *Aquatic Ecosystems and Global Climate Change* (Arlington, Va.: Pew Center on Global Climate Change, 2002).

24. P. C. D. Milly, K. A. Dunne, and A. V. Vecchia, "Global Pattern of Trends in Streamflow and Water Availability in a Changing Climate," *Nature* 438 (2006): 347–350.

25. B. D. Richter, R. Mathews, D. L. Harrison, and R. Wigington, "Ecologically Sustainable Water Management: Managing River Flows for Ecological Integrity," *Ecological Applications* 13 (2003): 206–224.

26. T. G. O'Brien and M. F. Kinnaird, "Caffeine and Conservation," *Science* 300 (2003): 587.

27. J. A. Foley, R. DeFries, G. P. Asner, C. Barford, G. Bonan, et al., "Global Consequences of Land Use," *Science* 309 (2005): 570–574.

28. M. Wackernagel, C. Monfreda, and D. Deumling, "Ecological Footprint of Nations November 2002 Update: How Much Nature Do They Use? How Much Nature Do They Have?" Redefining Progress Sustainability Issue Brief, November 2002 (available at: http://www.rprogress .org).

29. T. Dietz, E. A. Rosa, and R. York, "Driving the Human Ecological Footprint," *Frontiers in Ecology and the Environment* 5 (2007): 13–18.

30. E. W. Sanderson, M. Jaiteh, M. A. Levy, K. H. Redford, A. V. Wannebo, and G. Woolmer, "The Human Footprint and the Last of the Wild," *BioScience* 52 (2002): 891–904.

31. M. L. Imhoff, L. Bounoua, T. Ricketts, C. Loucks, R. Harriss, and W. T. Lawrence, "Global Patterns in Human Consumption of Net Primary Production," *Nature* 429 (2004): 870.

32. M. Wackernagel, N. B. Schulz, D. Deumling, A. Callejas Linares, M. Jenkins, et al., "Tracking the Ecological Overshoot of the Human Economy," *Proceedings of the National Academy of Sciences (US)* 99 (2002): 9266–9271.

33. Catton, *Overshoot*.

3. Shock Wave from the Insidious Explosion

1. W. R. Catton Jr., *Overshoot: The Ecological Basis of Revolutionary Change* (Urbana: University of Illinois Press, 1980).

2. J. Alroy, "A Multispecies Overkill Simulation of the End-Pleistocene Megafaunal Mass Extinction," *Science* 292 (2001): 1893–1896; R. C. Roverts, T. F. Flannery, L. K. Ayliffe, H. Yoshida, J. Olley, et al., "New Ages of the Last Australian Megafauna: Continent-Wide Extinction About 46,000 Years Ago," *Science* 292 (2001): 1888–1892.

3. G. H. Miller, M. L. Fogel, J. W. Magee, M. K. Gagan, S. J. Clarke, and B. J. Johnson, "Ecosystem Collapse in Pleistocene Australia and a Human Role in Megafaunal Extinction," *Science* 309 (2005): 287–289.

4. C. L. Redman, *Human Impact on Ancient Environments* (Tucson: University of Arizona Press, 1999).

5. T. L. Hunt, "Rethinking the Fall of Easter Island," *American Scientist* 94 (2006): 412–419.

6. P. R. Ehrlich, *Human Natures: Genes, Cultures, and the Human Prospect* (Washington, D.C.: Island Press, 2000).

7. R. W. Kates, B. L. Turner II, and W. C. Clark, "The Great Transformation," in B. L. Turner II, W. C. Clark, R. W. Kates, and J. F. Richards, eds., *The Earth as Transformed by Human Action: Global and Regional Changes in the Biosphere over the Past 300 Years* (Cambridge: Cambridge University Press, 1990), 1–17.

8. W. F. Ruddiman, *Plows, Plagues and Petroleum: How Humans Took Control of Climate* (Princeton, N.J.: Princeton University Press, 2005).

9. B. Moore III and B. H. Braswell Jr., "Planetary Metabolism: Understanding the Carbon Cycle," *Ambio* 23 (1994): 4–12.

10. P. Falkowski, R. J. Scholes, E. Boyle, J. Canadell, D. Canfield, et al. "The Global Carbon Cycle: A Test of Our Knowledge of Earth as a System," *Science* 290 (2000): 291–296.

11. W. H. Schlesinger, *Biogeochemistry: An Analysis of Global Change*, 2nd ed. (San Diego: Academic Press, 1997).

12. Kates et al., "Great Transformation."

13. E. Kolbert, *Field Notes from a Catastrophe: Man, Nature, and Climate Change* (New York: Bloomsbury, 2006).

14. T. Flannery, *The Weather Makers: How Man Is Changing the Climate and What It Means for Life on Earth* (New York: Atlantic Monthly Press, 2005).

15. "The Science of Climate Change," a joint statement issued by the Australian Academy of Sciences, Royal Flemish Academy of Belgium for Sciences, Royal Society of Canada, Caribbean Academy of Sciences, Chinese Academy of Sciences, French Academy of Sciences, German Academy of Natural Scientists Leopoldina, Indian National Science Academy, Indonesian Academy of Sciences, Royal Irish Academy, Accademia Nazionale dei Lincei (Italy), Royal Swedish Academy of Sciences, Turkish Academy of Sciences, and Royal Society (United Kingdom), *Science* 292 (2001): 1261.

16. T. Barker, I. Bashmakov, L. Bernstein, J. Bogner, P. Bosch, et al., "Intergovernmental Panel on Climate Change: Climate Change 2007— Summary for Policymakers" (available at: http://www.ipcc.ch/).

17. R. T. Watson, D. L. Albritton, T. Barker, I. A. Bashmakov, O. Canziani, et al., "Intergovernmental Panel on Climate Change: Climate Change 2001—Synthesis Report, Summary for Policymakers" (available at: http://www.ipcc.ch).

18. T. P. Barnett, D. W. Pierce, and R. Schnur, "Detection of Anthropogenic Climate Change in the World's Oceans," *Science* 292 (2001): 270–274.

19. S. Huang, H. N. Pollack, and P.-Y. Shen, "Temperature Trends over the Past Five Centuries Reconstructed from Borehole Temperatures," *Nature* 103 (2000): 756–758.

20. S. Levitus, J. I. Antonov, J. Wang, T. L. Delworth, K. W. Dixon, and A. J. Broccoli "Anthropogenic Warming of Earth's Climate System," *Science* 292 (2001): 267–270.

21. L. Hughes, "Biological Consequences of Global Warming: Is the Signal Already Apparent?" *TREE* 15 (2000): 56–61.

22. R. S. Bradley, M. Vuile, H. F. Diaz, and W. Vergara, "Threats to Water Supplies in the Tropical Andes," *Science* 312 (2006): 1755–1756.

23. S. C. B. Raper and R. J. Braithwaite, "Low Sea Level Rise Projections from Mountain Glaciers and Icecaps Under Global Warming," *Nature* 339 (2006): 311–314.

24. E. Rignot and P. Kanagaratnam, "Changes in the Velocity Structure of the Greenland Ice Sheet," *Science* 311 (2006): 986–990.

25. G. A. Meehl, W. M. Washington, W. D. Collins, J. M. Arblaster, A. Hu, et al., "How Much More Global Warming and Sea Level Rise?" *Science* 307 (2005): 1769–1772.

26. C. D. Thomas, A. Cameron, R. E. Green, M. Bakkenes, L. J. Beaumont, et al., "Extinction Risk from Climate Change," *Nature* 427 (2004): 145–148.

27. T. L. Root, J. T. Price, K. R. Hall, S. H. Schneider, C. Rosenzweig, and J. A. Pounds, "Fingerprints of Global Warming on Wild Animals and Plants," *Nature* 421 (2003): 57–60.

28. A. H. Fitter and R. S. R. Fitter, "Rapid Changes in Flowering Time in British Plants," *Science* 296 (2002): 1689–1691.

29. G. Beaugrand, P. C. Reid, F. Ibañez, J. A. Lindley, and M. Edwards, "Reorganization of North Atlantic Marine Copepod Biodiversity and Climate," *Science* 296 (2002): 1692–1694.

30. P. C. D. Milly, K. A. Dunne, and A. V. Vecchia, "Global Pattern of Trends in Streamflow and Water Availability in a Changing Climate," *Nature* 438 (2006): 347–350.

31. G. A. Meehl and C. Tebaldi, "More Intense, More Frequent, and Longer Lasting Heat Waves in the 21st Century," *Science* 305 (2004): 994–997.

32. B. Lomborg, *The Skeptical Environmentalist: Measuring the Real State of the World* (Cambridge: Cambridge University Press, 2001).

33. J. A. Patz, D. Campbell-Lendrum, T. Holloway, and J. A. Foley, "Impact of Regional Climate Change on Human Health," *Nature* 438 (2005): 310–317.

34. S. P. Leatherman, "Social and Economic Costs of Sea Level Rise," in B. Douglas, M. S. Kearney, and S. P. Leatherman, eds., *Sea Level Rise: History and Consequences* (San Diego: Academic Press, 2001), 181–223.

35. R. T. Watson, "Climate Change: The Political Situation," *Science* 302 (2003): 1925–1926.

36. A. L. Westerling, H. G. Hidalgo, D. R. Cayan, and T. W. Swetnam, "Warming and Earlier Spring Increase Western U.S. Forest Wildfire Activity," *Science* 313 (2006): 940–943.

37. C. D. Harvell, "Climate Warming and Disease Risks for Terrestrial and Marine Biota," *Science* 21 (2002): 2158–2162.

38. T. A. Gardner, I. M. Côté, J. A. Gill, A. Grant, and A. R. Watkinson, "Long-Term Region-Wide Declines in Caribbean Corals," *Science* 301 (2003): 958.

39. T. Goreau, T. McClanahan, R. Hayes, and A. Strong, "Conservation of Coral Reefs After the 1998 Global Bleaching Event," *Conservation Biology* 14 (2000): 5–15.

40. J. M. Pandolfi, R. H. Bradbury, E. Sala, T. P. Hughes, K. A. Bjorndal, et al., "Global Trajectories of the Long-Term Decline of Coral Reef Ecosystems," *Science* 301 (2003): 955.

41. T. P. Hughes, A. H. Baird, D. R. Bellwood, M. Card, S. R. Connolly, et al., "Climate Change, Human Impacts, and the Resilience of Coral Reefs," *Science* 301 (2003): 929–933.

42. J. C. Orr, V. J. Fabry, O. Aumont, L. Bopp, S. C. Doney, et al., "Anthropogenic Ocean Acidification over the Twenty-first Century and Its Impact on Calcifying Organisms," *Nature* 437 (2005): 681–686.

43. S. A. Zimov, E. A. G. Schuur, and F. S. Chapin III, "Permafrost and the Global Carbon Budget," *Science* 312 (2006): 1612–1613.

44. R. Wieder, "Past, Present, and Future Peatland Carbon Balance: An Empirical Model Based on ^{210}Pb-dated Cores," *Ecological Applications* 11 (2001): 327–342.

45. W. F. Laurance and G. B. Williamson, "Positive Feedbacks Among Forest Fragmentation, Drought, and Climate Change in the Amazon," *Conservation Biology* 15 (2001): 1529–1535.

46. Flannery, *Weather Makers*.

47. B. J. Peterson, R. M. Holmes, J. W. McClelland, C. J. Vörösmarty, R. B. Lammers, et al., "Increasing the River Discharge to the Arctic Ocean," *Science* 298 (2002): 2171.

48. T. R. Karl and K. E. Trenberth, "Modern Global Climate Change," *Science* 302 (2003): 1719–1723.

49. J. Lovelock, *Gaia: A New Look at Life on Earth* (Oxford: Oxford University Press, 1995).

50. R. W. Hahn and C. R. Sunstein, "The Precautionary Principle as a Basis for Decision Making," *Economists' Voice* 2 (2005): 1–9.

51. W. K. Stevens, "If Climate Changes, It May Change Quickly," *New York Times*, January 27, 1998, F1.

52. G. Yohe, N. Andronova, and M. Schlesinger, "To Hedge or Not Against an Uncertain Climate Future?" *Science* 306 (2004): 416.

53. S. Pacala and R. Socolow, "Stabilization Wedges: Solving the Climate Problem for the Next 50 Years with Current Technologies," *Science* 305 (2004): 968–972.

54. K. Warr, "The Ozone Layer," in P. Smith and K. Warr, eds., *Global Environmental Issues* (London: Hodder and Stoughton, 1991), 121–171.

55. T. K. Tromp, R.-L. Shia, M. Allen, J. M. Eiler, and Y. L. Yung, "Potential Environmental Impact of a Hydrogen Economy on the Stratosphere," *Science* 300 (2003): 1740–1742.

56. M. J. Newchurch, E.-S. Yang, D. N. Cunnold, G. C. Reinsel, J. M. Zawodny, and J. M. Russell III, "Evidence for Slowdown in Stratospheric Ozone Loss: First Stage of Ozone Recovery," *Journal of Geophysical Research* 108 (2003): 4507.

57. D. Pimentel, C. Harvey, P. Resosudarmo, K. Sinclair, D. Kurz, et al., "Environmental and Economic Costs of Soil Erosion and Conservation Benefits," *Science* 267 (1995): 1117–1123.

58. U.S. Environmental Protection Agency, "EPA National Water Quality Report," 1997 (available at: http://www.epa.gov/watrhome/resources).

59. P. M. Vitousek, J. Aber, R. W. Howarth, G. E. Likens, P. A. Matson, et al., "Human Alteration of the Global Nitrogen Cycle: Causes and Consequences," *Issues in Ecology* 1 (1997): 2–15.

60. J. Kaiser, "The Other Global Pollutant: Nitrogen Proves Tough to Curb," *Science* 294 (2001): 1268–1269.

61. M. Meybeck, "Carbon, Nitrogen, and Phosphorus Transport by World Rivers," *American Journal of Science* 282 (1982): 401–450.

62. V. H. Smith, "Eutrophication of Freshwater and Coastal Marine Ecosystems: A Global Problem," *Environmental Science and Pollution Research* 10 (2003): 126–139.

63. J. M. Burkholder and H. B. Glasgow Jr., "*Pfiesteria piscicida* and Other *Pfiesteria*-like Dinoflagellates: Behavior, Impacts, and Environmental Controls," *Limnology and Oceanography* 42 (1997): 1052–1075.

64. B. Underdal, O. Skulburg, E. Dahl, and T. Aune, "Disastrous Bloom of *Chrysochromulina polylepis* (Prymnesiophyceae) in Norwegian Coastal Waters: 1988 Mortality in Marine Biota," *Ambio* 18 (1989): 265–270; N. N. Rabalais, R. E. Turner, W. J. Wiseman Jr., and Q. Dortch, "Consequences of the 1993 Mississippi River Flood in the Gulf of Mexico," *Regulated Rivers Resource Management* 14 (1998): 161–177.

65. Burkholder and Glasgow, "*Pfiesteria piscicida* and Other *Pfiesteria*-like Dinoflagellates."

66. R. Dietz, M. Heid-Jorgensen, and T. Harkonen, "Mass Deaths of Harbor Seals (*Phoca vitulina*) in Europe," *Ambio* 18 (1989): 258–264.

67. P.M. Glibert, J. Harrison, C. Heil, and S. Seitzinger, "Escalating Worldwide Use of Urea: A Global Change Contributing to Coastal Eutrophication," *Biogeochemistry* 77 (2006): 441–463.

68. J.M. Pandolfi, J.B.C. Jackson, N. Baron, R.H. Bradbury, H.M. Guzman, et al., "Are U.S. Coral Reefs on the Slippery Slope to Slime?" *Science* 307 (2005): 1725–1726.

69. N.N. Rabalais, R.E. Turner, and D. Scavia, "Beyond Science into Policy: Gulf of Mexico Hypoxia and the Mississippi River," *BioScience* 52 (2002): 129–142.

70. C.S. Stow, S. Qian, and J.K. Craig, "Declining Threshold for Hypoxia in the Gulf of Mexico," *Environmental Science and Technology* 39 (2005): 716–723.

71. D. Scavia and S. Bricker, "Coastal Eutrophication in the United States," *Biogeochemistry* 79 (2006): 187–208.

72. G.D. Cooke, E.B. Welch, S.A. Peterson, and P.R. Newroth, *Restoration and Management of Lakes and Reservoirs*, 2nd ed. (Boca Raton, Fla.: Lewis, 1993).

73. B.J. Peterson, W.M. Wollheim, P.J. Mulholland, J.R. Webster, J.L. Meyer, et al., "Control of Nitrogen Export from Watersheds by Headwater Streams," *Science* 292 (2001): 86–90.

74. D. Pimentel, H. Acquay, M. Biltonen, P. Rice, M. Silva, et al., "Environmental and Economic Costs of Pesticide Use," *BioScience* 42 (1992): 750–760.

75. P.J. Daborn, J.L. Yen, M. R. Bogwitz, G. Le Goff, E. Feil, et al., "A Single p450 Allele Associated with Insecticide Resistance in *Drosophila*," *Science* 297 (2002): 2253–2256.

76. S.L. Simonich and R.A. Hites, "Global Distribution of Persistent Organochlorine Compounds," *Science* 269 (1995): 1851–1854.

77. F. Wania and D. Mackay, "Global Fractionation and Cold Condensation of Low Volatility Organochlorine Compounds in Polar Regions," *Ambio* 22 (1993): 10–18.

78. R. Renner, "Tracking the Dirty Byproducts of a World Trying to Stay Clean," *Science* 306 (2004): 1887.

79. R. Dietz, "Trends in Mercury in Hair of Greenlandic Polar Bears (*Ursus maritimus*) During 1892–2001," *Environmental Science and Technology* 40 (2006): 1120–1125.

80. P. Webster, "Study Finds Heavy Contamination Across Vast Russian Arctic," *Science* 306 (2004): 1875.

81. K. Kannan, J.C. Franson, W.W. Bowerman, K.J. Hansen, P.D. Jones, and J.P. Giesy, "Perfluorooctane Sulfonate in Fish-Eating Water Birds Including Bald Eagles and Albatrosses," *Environmental Science and Technology* 35 (2001): 3065–3070.

82. J. Bohannon, "A Toxic Odyssey," *Science* 304 (2004): 1584–1586.

83. J.M. Blais, D.W. Schindler, D.C.G. Muir, M. Sharp, D. Donald, et al., "Melting Glaciers: A Major Source of Persistent Organochlorines to Subalpine Bow Lake in Banff National Park, Canada," *Ambio* 30 (2001): 410–415.

84. L. Shen, F. Wania, Y.D. Lei, C. Teixeira, D.C.G. Muir, and T.F. Bidleman, "Atmospheric Distribution and Long-Range Transport Behavior of Organochlorine Pesticides in North America," *Environmental Science and Technology* 39 (2005): 409.

85. C. Davidson, H.B. Shaffer, and M.R. Jennings, "Declines of the California Red-Legged Frog: Climate, UV-B, Habitat, and Pesticides Hypothesis," *Ecological Applications* 11 (2001): 464–479.

86. C. Davidson, "Declining Downwind: Amphibian Population Declines in California and Historical Pesticide Use," *Ecological Applications* 14 (2004): 1892–1902.

87. K. Kidd, A.D.W. Schindler, D.C.G. Muir, W.L. Lockhart, and R.H. Hesslein, "High Concentrations of Toxaphene in Fishes from a Subarctic Lake," *Science* 269 (1995): 240–242.

88. D.W. Koplin, E.T. Furlong, M.T. Meyer, E.M. Thurman, S.D. Zaugg, et al., "Pharmaceuticals, Hormones, and Other Organic Wastewater Contaminants in U.S. Streams, 1999–2000: A National Reconnaissance," *Environmental Science and Technology* 36 (2002): 1202–1211.

89. P. Stahlschmidt-Allner, B. Allner, J. Römbke, and T. Knacker, "Endocrine Disrupters in the Aquatic Environment," *Environmental Science and Pollution Research* 4 (1997): 155–162; C. Sonnenschein and A. M. Soto, "An Updated Review of Environmental Estrogen and Androgen Mimics and Antagonists," *Journal of Steroid Biochemistry and Molecular Biology* 65 (1997): 43–150.

90. T.B. Hayes, "There Is No Denying This: Defusing the Confusion About Atrazine," *BioScience* 54 (2004): 1138–1149.

91. G. Blumenstyk, "The Price of Research," *Chronicle of Higher Education* 50 (2003): A26.

92. J.A. McLachlan and S.F. Arnold, "Environmental Estrogens," *American Scientist* 84 (1996): 452–461.

93. T. Colborn, D. Dumanoski, and J. P. Myers, *Our Stolen Future: Are We Threatening Our Fertility, Intelligence, and Survival?—A Scientific Detective Story* (New York: Dutton, 1996).

94. Ibid.

95. H. Tschäpe, "The Spread of Plasmids as a Function of Bacterial Adaptability," *FEMS Microbiology Ecology* 15 (1994): 23–32.

96. A. Mosier, M. Bleken, P. Chaiwanakupt, E. Ellis, J. Freney, et al., "Policy Implications of Human-Accelerated Nitrogen Cycling," *Biogeochemistry* 52 (2001): 281–320.

97. D. Tilman, J. Farione, B. Wolff, C. D'Antonio, A. Dobson, et al., "Forecasting Agriculturally Driven Global Environmental Change," *Science* 292 (2001): 281–284.

4. Weeds and Shrinking Violets

1. D. Simberloff, "Skewed Skepticism," *American Scientist* 90 (2002): 184–186.

2. L. Kaufman, "Catastrophic Change in Species-Rich Freshwater Ecosystems: The Lessons of Lake Victoria," *Bioscience* 42 (1992): 846–858.

3. W. P. Kovalak, G. D. Longton, and R. D. Smithee, "Infestation of Power Plant Water Systems by the Zebra Mussel (*Dreissena polymorpha* Pallas)," in T. F. Nalepa and D. W. Schloesser, eds., *Zebra Mussels: Biology, Impacts, and Control* (Boca Raton, Fla.: Lewis, 1993), 359–379; M. L. Ludyanskiy, D. McDonald, and D. MacNeill, "Impact of the Zebra Mussel, a Bivalve Invader," *BioScience* 43 (1993): 533–544.

4. D. D. Biesboer and N. Eckardt, *Euphorbia esula* (Arlington, Va.: Nature Conservancy, 2000).

5. D. A. Bangsund, F. L. Leistritz, and J. A. Leitch, *Predicted Future Economic Impacts of Biological Control of Leafy Spurge in the Upper Midwest*, NDSU Agricultural Economics Report, no. 382 (Fargo: North Dakota State University Agricultural Experiment Station, 1997).

6. D. Goulson, "Effects of Introduced Bees in Native Ecosystems," *Annual Review of Ecology and Systematics* 34 (2003): 1–26.

7. T. H. Fritts and D. Leasman-Tanner, "The Brown Treesnake on Guam: How the Arrival of One Invasive Species Damaged the Ecology, Commerce, Electrical Systems, and Human Health on Guam—A Comprehensive Information Source," 2001 (available at: http://111.aphis/usda/gov/).

8. J. Van Driesche and R. Van Driesche, *Nature Out of Place: Biological Invasions in the Golden Age* (Washington, D.C.: Island Press, 2000).

9. D. K. A. Barnes, "Invasion by Marine Life on Plastic Debris," *Nature* 416 (2002): 808–809.

10. P. W. Price, *Biological Evolution* (Fort Worth, Tex.: Saunders, 1996).

11. N. Meyer, "The Biodiversity Outlook: Endangered Species and Endangered Ideas," in J. Schogren and J. Tschirhart, eds., *Protecting Endangered Species in the United States: Biological Needs, Political Realities, Economic Choices* (Cambridge: Cambridge University Press, 2001), 138–165.

12. L. P. Koh, R. R. Dunn, N. S. Sodhi, R. K. Colwell, H. C. Proctor, and V. S. Smith, "Species Coextinctions and the Biodiversity Crisis," *Science* 305 (2004): 1632–1634.

13. S. N. Stuart, J. S. Chanson, N. A. Cox, B. E. Young, A. S. L. Rodrigues, et al., "Status and Trends of Amphibian Declines and Extinctions Worldwide," *Science* 306 (2004): 1783.

14. D. L. Strayer, "Challenges for Freshwater Invertebrate Conservation," *Journal of the North American Benthological Society* 25 (2006): 271–287.

15. J. A. Thomans, M. G. Telfer, D. B. Roy, C. D. Preston, J. J. D. Greenwood, et al., "Comparative Losses of British Butterflies, Birds, Plants and the Global Extinction Crisis," *Science* 303 (2004): 1879–1881.

16. J. Liu, G. Daily, P. R. Ehrlich, and G. W. Luck, "Effects of Household Dynamics on Resource Consumption and Biodiversity," *Nature* 421 (2003): 530–532.

17. G. Ceballos, P. R. Ehrlich, J. Soberón, I. Salazar, and J. P. Fay, "Global Mammal Conservation: What Must We Manage?" *Science* 309 (2005): 603–607.

18. M. R. Whiles, K. R. Lips, C. M. Pringle, S. S. Kilham, R. J. Bixby, et al., "The Effects of Amphibian Population Declines on the Structure and Function of Neotropical Stream Ecosystems," *Frontiers in Ecology* 4 (2006): 27–34.

19. J. R. Mendelson III, K. R. Lips, R. W. Gagliardo, G. B. Rabb, J. P. Collins, et al., "Confronting Amphibian Declines and Extinctions," *Science* 313 (2006): 48.

20. J. A. Pounds, M. R. Bustamante, L. A. Coloma, J. A. Consuegra, M. P. L. Fogden, et al., "Widespread Amphibian Extinctions from Epidemic Disease Driven by Global Warming," *Nature* 439 (2006): 161–167.

21. J. B. C. Jackson, M. X. Kirby, W. H. Berger, K. A. Bjorndal, L. W. Botsford, et al., "Historical Overfishing and the Recent Collapse of Coastal Ecosystems," *Science* 293 (2001): 629–638.

22. J. A. Estes, M. T. Tinker, T. M. Williams, and D. F. Doak, "Killer Whale Predation on Sea Otters Linking Oceanic and Nearshore Ecosystems," *Science* 282 (1998): 473–475.

23. R. M. May, "How Many Species Are There on Earth?" *Science* 241 (1988): 1441–1449.

24. M. E. Soulé and M. A. Sanjayan, "Conservation Targets: Do They Help?" *Science* 279 (1998): 2060–2061.

25. C. Roberts, C. J. McClean, J. E. N. Veron, J. P. Hawkins, G. R. Allen, et al., "Marine Biodiversity Hotspots and Conservation Priorities for Tropical Reefs," *Science* 295 (2002): 1280–1284.

26. J. B. Hughes, G. C. Daily, and P. R. Ehrlich, "Population Diversity: Its Extent and Extinction," *Science* 278 (1997): 689–692.

27. E. O. Wilson, *The Diversity of Life* (Cambridge, Mass.: Harvard University Press, 1992).

28. M. E. Gilpin and M. E. Soulé, "Minimal Viable Populations: Processes of Species Extinction," in M. E. Soulé, ed., *Conservation Biology: The Science of Scarcity and Diversity* (Sunderland, Mass.: Sinauer, 1986), 19–34.

29. Millennium Ecosystem Assessment, *Ecosystems and Human Well-Being: Synthesis* (Washington, D.C.: Island Press, 2005) (available at: http://ma.caudillweb.com/en/Products.Synthesis.aspx).

30. B. Commoner, *The Closing Circle: Nature, Man, and Technology* (New York: Knopf, 1971), 41.

5. Survival on a Finite Earth

1. R. Axelrod and D. Dion, "The Further Evolution of Cooperation," *Science* 242 (1988): 1385–1389.

2. G. T. Gardner and P. C. Stern, *Environmental Problems and Human Behavior* (Boston: Allyn and Bacon, 1996).

3. G. Hardin, "The Tragedy of the Commons," *Science* 162 (1968): 1243–1248.

4. S. Sharma, "Managing Environment: A Critique of 'The Tragedy of the Commons,'" *Journal of Human Ecology* 12 (2001): 1–9.

5. J. B. C. Jackson, M. X. Kirby, W. H. Berger, K. A. Bjorndal, L. W. Botsford, et al., "Historical Overfishing and the Recent Collapse of Coastal Ecosystems," *Science* 293 (2001): 629–638.

6. Gardner and Stern, *Environmental Problems and Human Behavior.*

7. Ibid.

8. E. Ostrom, T. Dietz, N. Dolšak, P. C. Stern, S. Stonich, and E. U. Weber, eds., *The Drama of the Commons* (Washington, D.C.: National Academy Press, 2002).

9. P. C. Stern, T. Dietz, E. Ostrom, N. Dolšak, and S. Stonich, "Knowledge and Questions After 15 Years of Research," in E. Ostrom, T. Dietz, N. Dolšak, P. C. Stern, S. Stonich, and E. U. Weber, eds., *The Drama of the Commons* (Washington, D.C.: National Academy Press, 2002), 445–486.

10. Gardner and Stern, *Environmental Problems and Human Behavior.*

11. W. D. Hamilton, "The Genetical Evolution of Social Behaviour I," *Journal of Theoretical Biology* 7 (1964): 1–16; R. L. Trivers, "The Evolution of Reciprocal Altruism," *Quarterly Review of Biology* 46 (1971): 35–57.

12. R. Dawkins, *The Selfish Gene* (Oxford: Oxford University Press, 1976).

13. E. Sober, *From a Biological Point of View: Essays in Evolutionary Philosophy* (Cambridge: Cambridge University Press, 1994).

14. S. H. Schwartz, "Normative Influences on Altruism," in L. Berkowitz, ed., *Advances in Experimental Social Psychology* (New York: Academic Press, 1977), 10:221–279.

15. Ibid.

16. P. J. Richerson, R. T. Boyd, and J. Henrich, "Cultural Evolution of Human Cooperation," in P. Hammerstein, ed., *Genetic and Cultural Evolution of Cooperation* (Cambridge, Mass.: MIT Press, 2003), 357–388.

17. K. Sigmund, *Games of Life: Explorations in Ecology, Evolution and Behaviour* (Oxford: Oxford University Press, 1993); W. E. Grant and P. B. Thompson, "Integrated Ecological Models: Simulation of Socio-cultural Constraints on Ecological Dynamics," *Ecological Modeling* 100 (1997): 43–59.

18. G. S. Becker, *The Economic Approach to Human Behavior* (Chicago: University of Chicago Press, 1976).

19. M. Kurlansky, *Cod: The Biography of the Fish That Changed the World* (New York: Penguin, 1997).

20. F. Fiegna, Y.-T. N. Yu, S. V. Kadam, and G. J. Velicer, "Evolution of an Obligate Social Cheater to a Superior Cooperator," *Nature* 441 (2006): 310–314.

21. Trivers, "Evolution of Reciprocal Altruism."

22. P. J. Deadman, "Modeling Individual Behaviour and Group Performance in an Intelligent Agent-Based Simulation of the Tragedy of the Commons," *Journal of Environmental Management* 56 (1999): 159–172.

23. S. F. Brosnan and B. M. de Waal, "Monkeys Reject Unequal Pay," *Nature* 425 (2003): 297–299; J. R. Stevens, "The Selfish Nature of Generosity:

Harassment and Food Sharing in Primates," *Proceedings of the Royal Society of London* 271 (2004): 451–456.

24. D. J. F. de Quervain, U. Fischbacher, V. Treyer, M. Schellhammer, U. Schnyder, et al., "The Neural Basis of Altruistic Punishment," *Science* 305 (2004): 1254–1258.

25. D. Knoch, A. Pascual-Leone, K. Meyer, V. Treyer, and E. Fehr, "Diminishing Reciprocal Fairness by Disrupting the Right Prefrontal Cortex," *Science* 314 (2006): 829–832.

26. E. Fehr and S. Gätcher, "Altruistic Punishment in Humans," *Nature* 415 (2002): 137–140.

27. K. Sigmund, E. Gehr, and A. Nowak, "The Economics of Fair Play," *Scientific American*, January 2002, 83–87.

28. J. Henrich, R. McElreath, A. Barr, J. Ensminger, C. Barrett, et al., "Costly Punishment Across Human Societies," *Science* 312 (2006): 1767–1770.

29. C. F. Camerer and E. Fehr, "When Does 'Economic Man' Dominate Social Behavior?" *Science* 311 (2006): 47–52.

30. T. R. Tyler and S. L. Blader, *Cooperation in Groups: Procedural Justice, Social Identity, and Behavioral Engagement* (Philadelphia: Taylor and Francis, 2000).

31. Camerer and Fehr, "When Does 'Economic Man' Dominate Social Behavior?"

32. M. A. Nowak and K. Sigmund, "Evolution of Indirect Reciprocity," *Nature* 437 (2005): 1291–1298.

33. N. Uphoff and J. Langholz, "Incentives for Avoiding the Tragedy of the Commons," *Environmental Conservation* 25 (1998): 251–261.

34. V. Corral-Verdugo, M. Frías-Armenta, F. Pérez-Urias, V. Orduña-Cabrera, and N. Espinoza-Gallego, "Residential Water Consumption, Motivation for Conserving Water and the Continuing Tragedy of the Commons," *Environmental Management* 30 (2002): 527–535.

35. J. Pretty, "Social Capital and the Collective Management of Resources," *Science* 302 (2003): 1912–1914.

36. L. E. Johnson, "Future Generations and Contemporary Ethics," *Environmental Values* 12 (2003): 471–497.

37. R. Costanza, "Social Traps and Environmental Policy," *BioScience* 37 (1987): 407–412.

38. R. S. Nickerson, *Psychology and Environmental Change* (Mahwah, N.J.: Erlbaum, 2003).

39. E. Fehr and B. Rochekbach, "Detrimental Effects of Sanctions on Human Altrusim," *Nature* 422 (2003): 137–140.

40. E. O. Wilson, *The Future of Life* (New York: Knopf, 2002).

41. S. J. DeCaneo, "Economic Analysis, Environmental Policy, and Intergenerational Justice in the Reagan Administration: The Case of the Montreal Protocol," *International Environmental Agreements: Politics, Law and Economics* 3 (2003): 299–321.

42. T. C. Haab and K. E. McConnell, "Social Norms and Illicit Behavior: An Evolutionary Model of Compliance," *Journal of Environmental Management* 66 (2002): 67–76.

43. D. W. Orr, *Earth in Mind: On Education, Environment, and the Human Prospect* (Washington, D.C.: Island Press, 1994).

6. Why Humans Foul the Nest

1. E. O. Wilson, *Sociobiology: The New Synthesis* (Cambridge, Mass.: Harvard University Press, 1975); M. Sahlins, *The Use and Abuse of Biology: An Anthropological Critique of Sociobiology* (Ann Arbor: University of Michigan Press, 1976).

2. A. Hüttermann, "What Is the Impact of a Philosophically Based Criticism of Socio-biology on the Scientific Community?" in P. Koslowski, ed., *Sociobiology and Bioeconomics: The Theory of Evolution in Biological and Economic Theory* (Berlin: Springer-Verlag, 1999), 183–196.

3. G. T. Gardner and P. C. Stern, *Environmental Problems and Human Behavior* (Boston: Allyn and Bacon, 1996); D. D. N. Winter and S. M. Koger, *The Psychology of Environmental Problems*, 2nd ed. (Mahwaw, N.J.: Erlbaum, 2004); R. S. Nickerson, *Psychology and Environmental Change* (Mahwaw, N.J.: Erlbaum, 2003).

4. E. O. Wilson, *Consilience: The Unity of Knowledge* (New York: Knopf, 1998); J. M. Diamond, *The Third Chimpanzee: The Evolution and Future of the Human Animal* (New York: HarperCollins, 1992).

5. P. R. Ehrlich, *Human Natures* (Washington, D.C.: Island Press, 2000).

6. E. W. Chu and J. R. Karr, "Environmental Impact, Concept and Measurement of," in S. A. Levin, ed., *Encyclopedia of Biodiversity* (San Diego: Academic Press, 2001), 2:557–577.

7. G. M. Hodgson, "Darwin, Veblen and the Problem of Causality in Economics," *History and Philosophy of the Life Sciences* 23 (2001): 385–423.

8. R. R. Nelson and S. G. Winter, "Evolutionary Theorizing in Economics," *Journal of Economic Perspectives* 16 (2002): 23–46.

9. D. Buller, *Adapting Minds: Evolutionary Psychology and the Persistent Quest for Human Nature* (Cambridge, Mass.: MIT Press, 2005).

10. L. A. Maguire, "What Can Decision Analysis Do for Invasive Species Management?" *Risk Analysis* 24 (2004): 859–868.

11. Y. Gilad, A. Oshlack, G. K. Smyth, T. P. Speed, and K. P. White, "Expression Profiling in Primates Reveals a Rapid Evolution of Human Transcription Factors," *Nature* 440 (2006): 242–245.

12. A. P. Hendry, "Rapid Evolution of Reproductive Isolation in the Wild: Evidence from Introduced Salmon," *Science* 290 (2000): 516–518.

13. R. H. Frank, *Passions Within Reason: The Strategic Role of the Emotions* (New York: Norton, 1988).

14. P. J. Richerson, R. Boyd, and P. Paciotti, "An Evolutionary Theory of Commons Management," in E. Ostrom, T. Dietz, N. Dolšak, P. C. Stern, S. Stonich, and E. U. Weber, eds., *The Drama of the Commons* (Washington, D.C.: National Academy Press, 2002), 403–445.

15. Gardner and Stern, *Environmental Problems and Human Behavior*.

16. W. R. Catton Jr., *Overshoot: The Ecological Basis of Revolutionary Change* (Urbana: University of Illinois Press, 1980).

17. M. Shermer, "The Soul of Science," *American Scientist* 93 (2005): 101–103.

18. World Wildlife Fund International, "Living Planet Report, 2006," ed. C. Hails (available at: http://www.footprintnetwork.org/newsletters/gfn_blast_0610.html).

19. E. O. Wilson, *Biophilia* (Cambridge, Mass.: Harvard University Press, 1984).

20. F. Fernández-Armesto, *Near a Thousand Tables: A History of Food* (New York: Free Press, 2003).

21. P. R. Ehrlich and A. H. Ehrlich, *One with Nineveh: Politics, Consumption, and the Human Future* (Washington, D.C.: Island Press, 2004).

22. R. H. Frank, *Luxury Fever: Why Money Fails to Satisfy in an Era of Excess* (New York: Free Press, 1999); J. Cairns Jr., "An Epic Struggle: Sustainability and the Emergence of a New Social Contract," *Social Contract* 9 (1999): 211–218.

23. E. Diener, O. Shigehiro, and R. E. Lucas, "Personality, Culture, and Subjective Well-Being: Emotional and Cognitive Evaluations of Life," *Annual Review of Psychology* 54 (2003): 403–425.

24. A. Martin, "Can Money Buy Happiness?" *Science* 268 (1995): 1113–1114.

25. Pew Research Center, "Are We Happy Yet? A Social Trends Report," 2006 (available at: http://pewresearch.org/pubs/301/are-we-happy-yet).

26. Winter and Koger, *Psychology of Environmental Problems*.

27. B.S. Low, *Why Sex Matters: A Darwinian Look at Human Behavior* (Princeton, N.J.: Princeton University Press, 2000).

28. Buller, *Adapting Minds*.

29. T. Zerjal, Y. Xue, G. Bertorelle, R. S. Wells, W. Bao, et al., "The Genetic Legacy of the Mongols," *American Journal of Human Genetics* 72 (2003): 717–721.

30. G. Miller, *The Mating Mind: How Sexual Choice Shaped the Evolution of Human Nature* (New York: Doubleday, 2000).

31. J.H. Barkow, *Darwin, Sex, and Status: Biological Approaches to Mind and Culture* (Toronto: University of Toronto Press, 1989).

32. Buller, *Adapting Minds*.

33. R.D. Alexander, *The Biology of Moral Systems* (New York: Aldine De Gruyter, 1987).

34. T. Veblen, *The Theory of the Leisure Class: An Economic Study of Institutions* (1899; repr., Mineola, N.Y.: Dover, 1994).

35. D.J. Futuyma, *Evolutionary Biology*, 3rd ed. (Sunderland, Mass.: Sinauer, 1987).

36. S. West, I. Pen, and A. Griffin, "Cooperation and Competition Between Relatives," *Science* 296 (2002): 72–75.

37. S. Bowles, "Group Competition, Reproductive Leveling, and the Evolution of Human Altruism," *Science* 314 (2006): 1569–1572.

38. M.A. Nowak, "Five Rules for the Evolution of Cooperation," *Science* 314 (2006): 1560–1563.

39. F. Werneken and M. Tomasello, "Altruistic Helping in Human Infants and Young Chimpanzees," *Science* 311 (2006): 1301–1303.

40. R.H. Frank, *Choosing the Right Pond: Human Behavior and the Quest for Status* (Oxford: Oxford University Press, 1985).

41. Wilson, *Biophilia*.

42. E.O. Wilson, *The Future of Life* (New York: Knopf, 2002).

43. C. Cokinos, *Hope Is the Thing with Feathers: A Personal Chronicle of Vanished Birds* (New York: Tarcher/Putnam, 2000).

44. M.E. Soulé, "Biophilia: The Unanswered Questions," in S.R. Kellert and E.O. Wilson, eds., *The Biophilia Hypothesis* (Washington, D.C.: Island Press, 1993), 441–455; J.M. Diamond, "New Guineans and Their Natural World," in S.R. Kellert and E.O. Wilson, eds., *The Biophilia Hypothesis* (Washington, D.C.: Island Press, 1993), 251–271.

45. O.R.W. Pergams and P.A. Zaradic, "Is Love of Nature in the U.S. Becoming Love of Electronic Media? 16-Year Downtrend in National Park Visits Explained by Watching Movies, Playing Video Games, Internet Use, and Oil Prices," *Journal of Environmental Management* 80 (2006): 387–393.

46. G. T. Seuss, *The Sneetches and Other Stories* (New York: Random House, 1961).

47. A. Agrawal, "Common Resources and Institutional Stability," in E. Ostrom, T. Deitz, N. Dolsak, P. C. Stern, S. Stonich, and E. U. Weber, eds., *The Drama of the Commons* (Washington, D.C.: National Academy Press, 2002), 41–85.

7. Searching for Answers

1. B. Lomborg, *The Skeptical Environmentalist: Measuring the Real State of the World* (Cambridge: Cambridge University Press, 2001).

2. R. W. Hahn and C. R. Sunstein, "The Precautionary Principle as a Basis for Decision Making," *Economists' Voice* 2 (2005): 1–9.

3. J. D. Marshall and M. W. Toffel, "Framing the Elusive Concept of Sustainability: A Sustainability Hierarchy," *Environmental Science and Technology* 39 (2005): 673.

4. H. E. Daly and J. B. Cobb Jr., *For the Common Good: Redirecting the Economy Toward Community, the Environment, and a Sustainable Future*, 2nd ed. (Boston: Beacon, 1994).

5. B. Commoner, *The Closing Circle: Nature, Man, and Technology* (New York: Knopf, 1971), 326.

6. N. Birdsall and S. W. Sinding, "How and Why Population Matters: New Findings, New Issues," in N. A. Birdsall, C. Kelley, and S. Sinding, eds., *Population Matters: Demographic Change, Economic Growth, and Poverty in the Developing World* (Oxford: Oxford University Press, 2001), 3–23.

7. P. Hawken, A. Lovins, and L. H. Lovins, *Natural Capitalism: Creating the Next Industrial Revolution* (Boston: Little, Brown, 1999).

8. R. Shinnar and F. Citro, "A Road Map to U.S. Decarbonization," *Science* 313 (2006): 1243–1244.

9. U.S. Department of Energy, Energy Information Administration, "Official Energy Statistics from the U.S. Government," n.d. (available at: http://www.eia.doe.gov/).

10. P. R. Ehrlich and A. H. Ehrlich, *One with Nineveh: Politics, Consumption, and the Human Future* (Washington, D.C.: Island Press, 2004).

11. P. R. Ehrlich, *The Population Bomb* (New York: Ballantine, 1968).

12. W. R. Catton Jr., *Overshoot: The Ecological Basis of Revolutionary Change* (Urbana: University of Illinois Press, 1980).

13. D. Pimentel, "Ethanol Fuels: Energy Balance, Economics, and Environmental Impacts Are Negative," *Natural Resources Research* 12 (2003): 127.

14. A. E. Farrell, R. J. Plevin, B. T. Turner, A. D. Jones, M. O'Hare, and D. M. Kammen, "Ethanol Can Contribute to Energy and Environmental Goals," *Science* 311 (2006): 506–508.

15. R. Hammerschlag, "Ethanol's Energy Return on Investment: A Survey of the Literature from 1990 to Present," *Environmental Science and Technology* 40 (2006): 1744–1750.

16. Interacademy Panel on International Issues, "Transition to Sustainability in the 21st Century: The Contribution of Science and Technology," 2000 (available at: http://www.interacademies.net/?id=3568).

17. Union of Concerned Scientists, "World Scientists' Warning to Humanity," 1992 (available at: http://www.ucsusa.org/ucs/about/1992-world-scientists-warning-to-humanity.html).

18. B. McKibben, *The End of Nature* (New York: Anchor, 1999).

19. Union of Concerned Scientists, "World Scientists' Call for Action at the Kyoto Climate Summit," 1997 (available at: http://www.ucsusa.org/global_warming/science/world-scientists-call-for-action-at-the-kyoto-climate-summit.html).

20. Food and Agriculture Organization of the United Nations, "Production of Cereals and Share in World," 2004 (available at: http://www.fao.org/statistics/yearbook/vol_1_1/pdf/b01.pdf).

21. J. E. Cohen and D. Tilman, "Biosphere 2 and Biodiversity: The Lessons So Far," *Science* 274 (1996): 1150–1151.

22. I. M. Côté and J. D. Reynolds, "Predictive Ecology to the Rescue?" *Science* 298 (2002): 1181–1182.

23. W. H. Schlesinger, "Global Change Ecology," *Trends in Ecology and Evolution* 21 (2006): 348–352.

24. A. Greeley, "Religion and Attitudes Toward the Environment," *Journal for the Scientific Study of Religion* 32 (1993): 19–28.

25. C. J. Dempsey and R. A. Butkus, *All Creation Is Groaning: An Interdisciplinary Vision for Life in a Sacred Universe* (Collegeville, Minn.: Liturgical Press, 1999).

26. R. M. Gross, "Toward a Buddhist Environmental Ethic," in H. Coward and D. C. Maguire, eds., *Visions of a New Earth: Religious Perspectives on Population, Consumption, and Ecology* (Albany: State University of New York Press, 2000), 147–160.

27. V. Nayaranan, "One Tree Is Equal to Ten Sons: Some Hindu Responses to the Problems of Ecology, Population, and Consumerism," in H. Coward and D. C. Maguire, eds., *Visions of a New Earth: Religious*

Perspectives on Population, Consumption, and Ecology (Albany: State University of New York Press, 2000), 111–146.

28. C. Yü, "Chinese Religions on Population, Consumption, and Ecology," in H. Coward and D. C. Maguire, eds., *Visions of a New Earth: Religious Perspectives on Population, Consumption, and Ecology* (Albany: State University of New York Press, 2000), 161–173; J. K. Olupona, "African Religions and the Global Issues of Population, Consumption, and Ecology," in H. Coward and D. C. Maguire, eds., *Visions of a New Earth: Religious Perspectives on Population, Consumption, and Ecology* (Albany: State University of New York Press, 2000), 175–199.

29. Gross, "Toward a Buddhist Environmental Ethic."

30. J. G. Speth, *Red Sky at Morning: America and the Crisis of the Global Environment* (New Haven, Conn.: Yale University Press, 2004).

31. G. T. Gardner and P. C. Stern, *Environmental Problems and Human Behavior* (Boston: Allyn and Bacon, 1996).

32. L. E. Johnson, "Future Generations and Contemporary Ethics," *Environmental Values* 12 (2003): 471–487.

33. M. E. Tucker and J. Grim, "Religions of the World and Ecology: Discovering the Common Ground," Forum on Religion and Ecology, n.d. (available at: http://environment.harvard.edu/religion/information/index .html).

34. International Parliament of World Religions, statement on "Global Ethics of Cooperation of Religions on Human and Environmental Issues" (available at: http://environment.harvard.edu/religion/information/index .html).

35. John Paul II and the Ecumenical Patriarch His Holiness Bartholomew I, "Common Declaration on Environmental Ethics," June 10, 2002 (available at: http://www.vatican.va/holy_father/john_paul_ii/ speeches/2002/june/documents/hf_jp-ii_spe_20020610_venice- declaration_en.html).

36. Quoted in Amanda Griscom Little, "Cizik Matters," October 5, 2005 (available at: http://www.grist.org/news/maindish/2005/10/05/cizik/).

37. S. R. Carpenter, D. Bolgrien, R. C. Lathrop, C. A. Stow, T. Reed, and M. A. Wilson, "Ecological and Economic Analysis of Lake Eutrophication by Nonpoint Pollution," *Australian Journal of Ecology* 23 (1998): 68–79.

38. D. D. N. Winter and S. M. Koger, *The Psychology of Environmental Problems*, 2nd ed. (Mahwah, N.J.: Erlbaum, 2004).

39. M. A. Wilson and S. R. Carpenter, "Economic Valuation of Freshwater Ecosystem Services in the United States: 1971–1997," *Ecological Applications* 9 (1999): 772–783.

40. R. Costanza, R. D'Arge, R. de Groot, S. Farber, M. Grasso, et al., "The Value of the World's Ecosystem Services and Natural Capital," *Nature* 387 (1997): 253–260.

41. N. E. Bockstael, A.M. Freeman, R. J. Kopp, P. R. Portnoy, and V. K. Smith, "On Measuring Economic Values for Nature," *Environmental Science and Technology* 34 (2000): 1384–1389.

42. K.-G. Mäler and J. R. Vincent, eds., *Handbook of Environmental Economics*, 3 vols. (Amsterdam: Elsevier, 2003).

43. P. Bohm, "Experimental Evaluations of Policy Instruments," in K.-G. Mäler and J. R. Vincent, eds., *Handbook of Environmental Economics*, vol. 1, *Environmental Degradation and Institutional Responses* (Amsterdam: Elsevier, 2003), 436–460.

44. A. Balmford, A. Bruner, P. Cooper, R. Costanza, S. Farber, et al., "Economic Reasons for Conserving Wild Nature," *Science* 297 (2002): 950–953.

45. National Research Council, *Valuing Ecosystem Services: Toward Better Environmental Decision-Making* (Washington, D.C.: National Academies Press, 2004).

46. G. Romp, *Game Theory: Introduction and Applications* (Oxford: Oxford University Press, 1997).

47. D. McKenzie-Mohr, "Promoting Sustainable Behavior: An Introduction to Community-Based Social Marketing," *Journal of Social Issues* 56 (2000): 543–554.

48. P. C. Stern, "Toward a Coherent Theory of Environmentally Significant Behavior," *Journal of Social Issues* 56 (2000): 407–424.

49. D. Hicks and M. A. Gwynne, *Cultural Anthropology* (New York: HarperCollins, 1996).

50. S. Oskamp, "Psychological Contributions to Achieving an Ecologically Sustainable Future for Humanity," *Journal of Social Issues* 56 (2000): 373–390.

51. Winter and Koger, *Psychology of Environmental Problems*.

52. Gardner and Stern, *Environmental Problems and Human Behavior*.

53. E. Ostrom, *Governing the Commons: The Evolution of Institutions for Collective Action* (Cambridge: Cambridge University Press, 1990).

54. J. T. Cacioppo, P. S. Visser, and C. L. Pickett, eds., *Social Neuroscience: People Thinking About People* (Cambridge, Mass.: MIT Press, 2006).

55. J. Diamond, *Collapse: How Societies Choose to Fail or Succeed* (New York: Viking, 2005).

56. P. Barnes, *Capitalism 3.0: A Guide to Reclaiming the Commons* (San Francisco: Berrett-Koehler, 2006).

57. M. O'Connor, *Is Capitalism Sustainable? Political Economy and the Politics of Ecology* (New York: Guilford, 1994).

58. B. A. Forster, *The Acid Rain Debate: Science and Special Interests in Policy Formation* (Ames: Iowa State University Press, 1993).

59. K. Palmer, W. E. Oates, and P. R. Portney, "Tightening Environmental Standards: The Benefit-Cost or No-Cost Paradigm?" *Journal of Economic Perspectives* 9 (1995): 119–132.

60. "The Greening of General Electric" (special report), *Economist*, December 10, 2005, pp. 77–79.

61. K. Peattie, *Environmental Marketing Management: Meeting the Green Challenge* (London: Pitman, 1995).

62. R. N. Stavins, "Experience with Market-Based Environmental Policy Instruments," in K.-G. Mäler and J. R. Vincent, eds., *Handbook of Environmental Economics*, vol. 1, *Environmental Degradation and Institutional Responses* (Amsterdam: Elsevier, 2003), 353–435.

63. R. H. Frank, *Passions Within Reason: The Strategic Role of the Emotions* (New York: Norton, 1988).

64. M. Hertsgaard, *Earth Odyssey: Around the World in Search of Our Environmental Future* (New York: Broadway Books, 1998).

65. R. E. Grumbine, "China's Emergence and the Prospects for Global Sustainability," *BioScience* 57 (2007): 249–255.

66. L. R. Brown, *Plan B: Rescuing a Planet Under Stress and a Civilization in Trouble* (New York: Norton, 2003).

67. P. Jepson, J. K. Jarvie, K. MacKinnon, and K. A. Monk, "The End for Indonesia's Forests?" *Science* 292 (2001): 859–861.

68. A. Gore, *Earth in the Balance: Ecology and the Human Spirit* (New York: Houghton Mifflin, 1992), 000.

8. No More Business as Usual

1. W. Allen, "My Speech to the Graduates," in *Side Effects* (New York: Random House, 1980), 79–85, at 81.

2. J. Diamond, *Collapse: How Societies Choose to Fail or Succeed* (New York: Viking, 2005).

3. E. O. Wilson, *The Future of Life* (New York: Knopf, 2002).

4. A. Edgar and P. Sedgwick, *Key Concepts in Cultural Theory* (London: Routledge, 1999).

5. Earth Charter Initiative, "The Earth Charter" (available at: http://www.earthcharter.org).

6. M. D. Hauser, *Moral Minds: How Nature Designed Our Universal Sense of Right and Wrong* (New York: HarperCollins, 2006).

7. E. O. Wilson, *On Human Nature* (Cambridge, Mass.: Harvard University Press, 1978).

8. J. Lopreato, *Human Nature and Biocultural Evolution* (Boston: Allen and Unwin, 1984).

9. S. M. De Bruyn, *Economic Growth and the Environment: An Empirical Analysis* (The Hague: Kluwer Academic, 2000).

10. Quoted in P. H. Raven, "Science, Sustainability, and the Human Prospect," *Science* 297 (2002): 954–958, at 955.

11. B. R. Copeland and M. S. Taylor, *Trade and the Environment: Theory and Evidence* (Princeton, N.J.: Princeton University Press, 2003).

12. P. Dasgupta, "Is Contemporary Economic Development Sustainable?" *Ambio* 31 (2002): 269–271.

13. D. E. Fisher, *Fire and Ice: The Greenhouse Effect, Ozone Depletion, and Nuclear Winter* (New York: Harper & Row, 1990).

14. C. Sagan, "The Atmospheric Consequences of Nuclear War," in P. Ehrlich, C. Sagan, D. Kennedy, and W. O. Roberts, eds., *The Nuclear Winter: The Cold and the Dark* (London: Sidgwick and Jackson, 1984), 1–39.

15. A. Robock, "The Climatic Aftermath," *Science* 295 (2002): 1242–1244.

16. J. Schell, *The Fate of the Earth* (New York: Knopf, 1982).

17. CBS News Poll, *New York Times*, February 22–26, 2006.

18. R. S. Nickerson, *Psychology and Environmental Change* (Mahwah, N.J.: Erlbaum, 2003).

19. I. Serageldin, "World Poverty and Hunger: The Challenge for Science," *Science* 296 (2002): 54–57.

20. U.S. Department of Commerce, Bureau of Economic Analysis, "Gross-Domestic-Product-by-Industry Accounts Intermediate Inputs by Industry," press release, April 24, 2007 (available at: http://www.bea.gov).

21. P. Jha, A. Mills, K. Hanson, L. Kumaranayake, L. Conteh, et al., "Improving the Health of the Global Poor," *Science* 295 (2002): 2036–2039.

22. United Nations World Food Program, "Annual Report, 2005" (available at: http://www.wfp.org).

23. D. D. N. Winter and S. M. Koger. *The Psychology of Environmental Problems*, 2nd ed. (Mahwah, N.J.: Erlbaum, 2004).

24. S. Kaplan, "Human Nature and Environmentally Responsible Behavior," *Journal of Social Issues* 56 (2000): 491–508.

25. D. M. Reiner, T. E. Curry, M. A. Defigueiredo, H. J. Herzog, S. D. Ansolabehere, et al., "American Exceptionalism? Similarities and Differ-

ences in National Attitudes Toward Energy Policy and Global Warming," *Environmental Science and Technology* 40 (2006): 2093–2098, at 2095.

26. G. T. Gardner and P. C. Stern, *Environmental Problems and Human Behavior* (Boston: Allyn and Bacon, 1996).

27. Diamond, *Collapse*.

28. M. Jenkins, "Prospects for Biodiversity," *Science* 302 (2003): 1175–1177.

29. P. R. Ehrlich and A. H. Ehrlich, *One with Nineveh: Politics, Consumption, and the Human Future* (Washington, D.C.: Island Press, 2004).

9. Consilience

1. E. O. Wilson, *Consilience: The Unity of Knowledge* (New York: Knopf, 1998).

2. S. R. Carpenter and C. Folke, "Ecology for Transformation," *Trends in Ecology and Evolution* 21 (2006): 309–315.

3. M. Palmer, E. Bernhardt, E. Chornesky, S. Collins, A. Dobson, et al., "Ecology for a Crowded Planet," *Science* 304 (2004): 1251–1252.

4. J. Cairns, "Ecological Restoration: A Major Component of Sustainable Use of the Planet," *Renewable Resources Journal* 17 (1999): 6–10.

5. International Association for Landscape Ecology, "What Is Landscape Ecology?" Welcome at the IALE Website (available at: http://www.landscape-ecology.org).

6. Earth Simulator Center, Japan Agency for Marine-Earth Science and Technology, n.d. (available at: http://www.es.jamstec.go.jp/index.en.html).

7. A. Balmford and W. Bond, "Trends in the State of Nature and Their Implications for Human Well-Being," *Ecology Letters* 8 (2005): 1218–1234.

8. M. Hsu, M. Bhatt, R. Adolphs, D. Tranel, and C. F. Camerer, "Neural Systems Responding to Degrees of Uncertainty in Human Decision-Making," *Science* 310 (2005): 1680.

9. C. F. Camerer and E. Fehr, "When Does 'Economic Man' Dominate Social Behavior?" *Science* 311 (2006): 47–52.

10. M. Burtsev and P. Turchin, "Evolution of Cooperative Strategies from First Principles," *Nature* 440 (2006): 1041–1044.

11. R. R. Nelson and S. G. Winter, "Evolutionary Theorizing in Economics," *Journal of Economic Perspectives* 16 (2002): 23–46.

12. P. Hammerstein and E. H. Hagen, "The Second Wave of Evolutionary Economics in Biology," *Trends in Ecology and Evolution* 20 (2005): 604–609.

13. B. Loken, "Consumer Psychology: Categorization, Inferences, Affect, and Persuasion," *Annual Reviews of Psychology* 57 (2006): 453–485.

14. T. Brader, *Campaigning for Hearts and Minds: How Emotional Appeals in Political Ads Work* (Chicago: University of Chicago Press, 2006).

15. G. T. Gardner and P. C. Stern, *Environmental Problems and Human Behavior* (Boston: Allyn and Bacon, 1996).

16. R. S. Nickerson, *Psychology and Environmental Change* (Mahwah, N.J.: Erlbaum, 2003).

17. Gardner and Stern, *Environmental Problems and Human Behavior.*

18. Nickerson, *Psychology and Environmental Change.*

19. E. Diener, S. Oishi, and R. E. Lucas, "Personality, Culture, and Subjective Well-Being: Emotional and Cognitive Evaluations of Life," *Annual Reviews of Psychology* 54 (2003): 403–425.

20. H. Daly and J. Farley, *Ecological Economics: Principles and Applications* (Washington, D.C.: Island Press, 2004).

21. R. Costanza, J. Cumberland, H. Daley, R. Goodland, and R. Norgaard, *Introduction to Ecological Economics* (Boca Raton, Fla.: St. Lucie Press, 1997).

22. H. E. Daly and J. B. Cobb Jr., *For the Common Good: Redirecting the Economy Toward Community, the Environment, and a Sustainable Future*, 2nd ed. (Boston: Beacon, 1994).

23. D. Schröter, W. Cramer, R. Leemans, I. C. Prentice, M. B. Araújo, et al., "Ecosystem Service Supply and Vulnerability to Global Change in Europe," *Science* 310 (2005): 1333–1337.

24. L. H. Goulder and R. N. Stavins, "An Eye on the Future," *Nature* 419 (2002): 673–674.

25. U.S. Society for Ecological Economics, "USSEE" (available at: http://www.ussee.org).

26. Daly and Farley, *Ecological Economics.*

27. R. Costanza, "Ecological Economics: Creating a Transdisciplinary Science," in P. H. May and R. Serôa da Motta, eds., *Pricing the Planet: Economic Analysis for Sustainable Development* (New York: Columbia University Press, 1996), 139–169.

28. S. Farber, R. Costanza, D. L. Childers, J. Erickson, K. Gross, et al., "Linking Ecology and Economics for Ecosystem Management," *BioScience* 56 (2006): 121–133.

29. P. H. May, "Sustainability: Challenges to Economic Analysis and Policy," in P. H. May and R. Serôa da Motta, eds., *Pricing the Planet: Economic Analysis for Sustainable Development* (New York: Columbia University Press, 1996), 12–28.

30. C. S. Pearson, *Global Economics and the Environment* (Cambridge: Cambridge University Press, 2000).

31. Daly and Farley, *Ecological Economics.*

32. Ibid.

33. N. Myers and J. Kent, *Perverse Subsidies: Tax $s Undercutting Our Economies and Environments Alike* (Winnipeg: International Institute for Sustainable Development, 1998).

34. P. Bartelmus, "Green Accounting for Sustainable Development," in P. H. May and R. Serôa da Motta, eds., *Pricing the Planet: Economic Analysis for Sustainable Development* (New York: Columbia University Press, 1996), 180–196.

35. J. Kaiser, "Turning Engineers into Resource Accountants," *Science* 285 (1999): 685–686.

36. Centers for Disease Control and Prevention, National Center for Health Statistics, "Health, United States, with Chartbook on Trends in Health of Americans," 2005 (available at: http://www.cdc.gov/nchs/data/hus/hus05.pdf#summary).

37. Nickerson, *Psychology and Environmental Change.*

38. Gardner and Stern, *Environmental Problems and Human Behavior.*

39. Ibid.

40. Biomimicry Institute, "Bat Inspired Walking Cane," case studies (available at: http://www.biomimicry.net/casestudiesB.htm).

41. P. R. Ehrlich and D. Kennedy, "Millennium Assessment of Human Behavior," *Science* 309 (2005): 562–563.

42. P. Barnes, *Capitalism 3.0: A Guide to Reclaiming the Commons* (San Francisco: Berrett-Koehler, 2006).

43. D. D. Doniger, A. V. Herzog, and D. A. Lashof, "An Ambitious, Centrist Approach to Global Warming Legislation," *Science* 314 (2006): 764–765.

44. E. O. Wilson, *The Future of Life* (New York: Knopf, 2002).

45. Millennium Ecosystem Assessment, *Ecosystems and Human Well-Being: Synthesis* (Washington, D.C.: Island Press, 2005).

46. The Dalai Lama, *The Universe in a Single Atom: The Convergence of Science and Spirituality* (New York: Morgan Road Books, 2005), 201.

Index

Numbers in italics refer to figures.

accountability, 81, 209, 212
acid precipitation (acid rain), 19, 164
agriculture, 36, 47, 51, 54, 58, 61, 70, 112,
    115, 119, 131, 180; antibiotic use in,
    60–61; best management practices in,
    120, 141; chemical use in, 36, 55–56,
    60–61; crops, 23, 26–27, 29, 33, 38, 42,
    48, 51, 58, 61, 148–150, 160, 179, 182;
    fertilizer runoff from, 51, 53–54;
    fertilizer use in, 27, 51–55, 52, 61, 64,
    81, 141, 143, 148–150, 210, 219–220;
    food production, 17, 23, 27, 29, 31, 42,
    52, 55, 120, 143, 150–151, 182; green
    revolution in, 27, 150; and habitat
    destruction, 61, 70, 104; irrigation in,
    23–24, 27, 31, 143, 150; sedimentation
    from, 51, 168; sustainable, 86–87, 102,
    106, 163; and water quality, 51–52, 55,
    196. See also chemicals; DDT; erosion;
    nitrogen; pesticides; species,
    extinction of; water
air pollution. See atmosphere:
    pollution of
Alexandrowicz, Ra'anan, 1

algae, 36, 53–54, 73, 75; toxic (algal
    blooms), 53–54
Allen, Woody, 171, 185
altruism. See human behavior:
    altruism as
American Geophysical Union, 40
amphibians, 71; and pesticides, 54.
    See also frogs; species: extinction of
Amundsen, Roald, 1
Anasazi, 15, 154
Antarctica, 3, 41, 42, 49, 69
anthropology, x, 160
antibiotics, 59, 61, 113, 141, 175, 185
Arctic, 1–3, 2, 23, 41–42, 46, 49, 57–59, 69,
    141
Arctic Ocean, 1, 46, 47
Aristotle, 85, 121
Arrhenius, Scante, 40
Asia, 1, 7, 32, 35, 46, 73, 115, 124
Atlantic Ocean, 47, 66
atmosphere, 4–5, 25, 36–40, 46–51, 57,
    103, 179–180, 195; pollution of, 29, 32,
    36–40, 58, 81, 84–85, 150, 176–177, 188,
    225–227; temperature of, 41–43, 103,

atmosphere (*continued*)
179–180. *See also* carbon dioxide; global warming; greenhouse gases; ozone
Australia, 7, 16, 35, 50, 53, 66, 68
automobiles/vehicles, 19, 143, 150, 175, 181–182, 200, 225, 227

Barlow, John Perry, 108
Barnes, Peter, 163, 208–209
behavioral science, 147, 197, 199, 206. *See also* human behavior
biochemicals, 5
biodiversity 9, 64, 68, 70, 76–78, 141, 193, 210. *See also* deforestation/logging; habitats: protection of; species: diversity of
biofuels. *See* ethanol
biologists, 76, 83, 105, 108, 192, 198, 203
biology, 7, 42, 152, 198
biomimicry, 207
biophilia, 117–118, 131
biosphere, 5, 60–61, 72, 111, 127, 135, 151, 156, 167
bioterrorism, 6
biotic features, 3
birds, 36, 56, 64–65, 68, 70–71, 71, 94, 124–125, 128, 162. *See also* species: extinction of
birth control, 112, 118, 123, 133, 145, 211. *See also* human population
blue-baby syndrome (methemoglobin-emia), 52
Broecker, Wallace, 34, 47
brown tree snakes, 68–69

Canada, 1, 7, 53, 58, 142, 166
capitalism, 163–164, 166
carbon dioxide (CO_2), 17, 26, 37, 37–40, 39, 64, 85, 103, 107, 217. *See also* atmosphere; deforestation/logging; global warming; greenhouse gases
Center for the Advancement of the Steady State Economy, 202
CFCs (chlorine- and fluorine-containing organic compounds), 49, 51, 104, 141, 176, 179, 210. *See also* ozone: depletion of
cheating. *See* human behavior: cheating as

chemicals, 57–61, 81, 85, 138–139, 141, 150, 160, 210; toxic (waste), 3, 31, 36, 53, 56–61, 81, 85, 138–139, 141, 150, 160, 210. *See also* agriculture: chemical use in; agriculture: fertilizer use in; pesticides
China, 1, 15–16, 24, 49, 57, 103, 112, 116, 145, 163, 167, 176, 179
climate, 5, 19, 23, 47, 149, 152, 206; computer models of, 42–43, 195
climate change, 1–3, 5, 19, 38, 40, 45–48, 78, 80, 84, 162, 209. *See also* global warming
Commoner, Barry, 140
computer modeling, 192–195
conservation. *See* energy: conservation of; environment/environmental; resources: conservation of; species: conservation of; water: conservation of
conservation biology, 191–193
consilience, 189–191, 202–203, 205, 207, 209
cooperation. *See* human behavior: cooperation as
corn, 21. *See also* ethanol
corporate behavior, 164–165
Costanza, Robert, 157, 159
cropland, 17, 24, 27, 28, 29–31, 52, 143, 148, 150–151, 180, 186–188, 217–218. *See also* agriculture
cultural anthropology, 160–161
cultural values, 17, 23, 159
cyclones. *See* weather: cyclones

Dalai Lama, 214
Daly, Herman, 140, 201–202
Dawkins, Richard, 89, 114
DDT (dichloro-diphenyl-trichloroethane), 56–60, 157. *See also* pesticides
"Dead Zone" (Gulf of Mexico), 54
deforestation/logging, 29, 34–36, 38, 40, 45–46, 76–78, 77, 80–81, 85, 120, 158, 168, 187, 227
democracy, 166
developed countries, 7, 16, 19, 53, 56, 60, 130, 134, 158, 176, 185; population growth in, 118; resource use in, 108,

125, 141–143, 186; standard of living in, 21, 30, 152, 177

developing countries, 7–8, 26, 29, 45, 56, 104, 130, 158, 178; population growth in, 16, 118, 123; resource use in, 186; standard of living in, 16–17, 88, 119, 177

Diamond, Jared, 109, 162, 172,

diseases, human, ix, 6, 14, 16, 42, 44, 60–65, 78, 113, 118, 120, 126, 131, 147, 152, 157–158, 176, 182, 186, 209, 211; AIDS, 79–80, 149–150, 233; avian flu, 65, 79; bubonic plague (Black Death), 64; Lyme disease, 80; malaria, 45, 80; Nipah virus, 79; West Nile virus, 64

displacement (behavior), 183

Dr. Suess (Theodor Seuss Geisel), 134

"Earth Charter, The" (Earth Charter Initiative), 173

Easter Island, 15, 35–36, 126

ecological: damage, 7–8, 29, 191; holocaust, 63, 68. See also ecosystems; footprint, ecological

ecology, 7, 106, 108, 151–152, 191, 196, 201; conservation, 157; industrial, 204; landscape, restoration, and urban, 191–193

economic/economics, 110, 147, 156, 160, 163, 190, 198, 201–203; activity, 17, 19, 30, 32, 142–143, 156, 187, 189, 203–204; cost-benefit analyses in, 198, 202; development, 26, 167, 177; discounting value in, 202; ecological, 157, 201–203; and environmental externalities (costs), 7, 106, 156, 160, 165, 202–203, 210–211; expansion, 17, 18, 19; global, 3, 38, 156; global total gross national product, 203–204; growth, 23, 201–202; health, 201, 203, 210; and human behavior, 92, 203; incentives, 160; microeconomics, 204; output, 18, 23; stability, 166; steady-state, 106, 201–203; sustainable, 202. See also ecosystems: goods and services in, value of

ecosystems, 3–4, 10, 35, 64, 71–73, 80, 100, 140, 151–152, 158, 179; aquatic, 27, 42, 73, 75, 79; function of, 43, 64, 141; goods and services in, value of,

23–24, 84, 101, 146, 156–160, 164, 170, 178, 199, 202–204, 208–210; health of, 9, 52, 61, 73, 80, 137, 178, 196, 209–210, 212

education, 32, 99–100, 102–108, 112, 134–135, 145–146, 168, 172, 183–184, 188, 190, 200, 205–207, 209, 211–214

Ehrlich, Paul, 12, 208

emission credit trading, 146, 158

energy, 3, 17, 32, 84, 149–150, 186–187; alternative sources of, 151; coal as source of, 19, 30, 37–39, 80, 186; conservation of, 200, 210; consumption (use) of, 4, 17, 19, 20, 21, 25–27, 30–31, 38–39, 39, 48, 103, 142–144, 148, 181, 189, 204; efficiency of, 120, 134, 141–144, 147–148, 153, 166, 200, 206–207, 215–216; ethanol as source of, 148–149, 198; fossil fuels as source of, 30, 33, 38, 40–41, 103, 120, 134, 148–150, 186; gasoline/oil as source of, 3–4, 11, 19, 20, 30–31, 37–39, 49, 101, 132, 134, 143, 148, 151, 157, 181, 186, 200, 203, 213, 215–216, 225, 227; hydroelectric, 24–26, 30–31; methane (natural gas) as source of, 46; nuclear, 30–31, 154, 186; production of, 149, 164, 186; solar, 25, 30, 40; wind as source of, 30; wood as source of, 5, 19, 36, 37, 39, 77, 80, 168, 171. See also deforestation/logging; forests

enlightenment, 171–172, 174, 188, 228; behavioral, 181; environmental, 172–176, 178; technological, 175

environment/environmental, 3, 8, 34, 35–36, 62, 86, 95, 108, 132–133, 137, 140, 162, 167, 171, 178, 182, 191, 205, 208; capacity of, to sustain human impacts, 3, 5, 10, 14, 17, 32, 33, 35–36, 84, 115–116, 134, 136, 151–152, 158, 172–173, 187–188; computer modeling of, 195; conservation of, 118, 130, 161; costs, 7, 80, 146, 149–150, 156, 159, 211; damage to, 34–82, 89–90, 102, 106, 115, 139, 148–149, 154, 157, 162, 165–166, 169, 171, 177, 179, 200, 202, 203, 212; degradation of, 7, 51, 63, 68, 79, 130, 155, 170, 183, 185, 199; education about, 105–106, 172, 188, 205–206; externalities,

environment/environmental (*continued*)
160, 210; global change in, 9, 64, 153,
184, 212; human impacts on, 3–13, 16,
18–82, 90, 100, 107–116, 119, 127,
130–131, 136, 138, 141–147, 150–151,
163–165, 172, 175, 177, 181, 184, 187–189,
203, 210, 213; issues, 3, 8, 51, 82–83,
102, 105, 108, 110, 135, 140, 153,
160–161, 166, 178, 184, 190, 197, 203,
207, 212; organizations, 167–169, 191;
pesticides in, 56–60; policies
(regulations) on, 160, 164–166, 192,
201–202, 210; threats to, 9, 81,
178–179; trends, xi, 8, 34–82, 109, 118,
185; values of, 84, 170
environmental science, 84, 106, 152,
190–191, 193
erosion, 5, 31, 51, 66, 149, 171; wind as
cause of, 29, 58. *See also* agriculture:
sedimentation from
ethanol, 148–149, 182, 198
Europe, 65–66, 142
evolution, 12, 68, 78, 85, 89, 92, 98,
109–110, 119, 152, 172, 174–175, 178,
190; biological and cultural, 85,
89–90, 104, 107, 109, 112–116, 117, 126,
129, 167, 173–174, 198, 205; and genes,
112–114, 118, 123–128, 174, 193; and kin
selection, 127–129; and mutation, 48,
113; natural selection as motor of, 40,
112–114, 118, 123–128, 130, 172–173;
and reproduction (sexual selection),
118, 125–126; technological, 110;
variation in, 112–113. *See also* human
behavior
externalities. *See* economic/economics:
environmental externalities
extinction. *See* species: extinction of

fertility, 15–16, 60, 119, 145, 147, 211
fertilizers. *See* agriculture: fertilizer
runoff from; agriculture: fertilizer
use in; chemicals; environment/
environmental; nitrogen
fish, 43, 54, 65, 70, 73, 74, 75, 94, 162, 196;
in Lake Victoria, 65. *See also* species:
extinction of
food, 5–6, 14, 23, 29, 32, 42, 48, 51, 54, 75,
93–94, 115, 118–120, 125, 139, 145, 150,

160, 175, 186; demand for, 29, 55, 61;
and malnutrition, 6, 26, 79, 182;
processed, 213; production of, 17, 27,
29, 31, 42, 52, 55, 143, 150–151, 182. *See
also* agriculture; cropland
food webs, 65, 73, 74, 75, 91
footprint, ecological, ix, 3, 11, 32–33, 88,
100, 116, 177, 202, 209–210, 212–213.
See also ecological: damage;
environment/environmental
forests, 17, 38, 39, 73, 77, 77, 81, 84, 158, 168,
185, 188, 196; and greenhouse gases,
38, 77, 158; rain forests, 29, 46, 73,
76–77, 141, 157, 168. *See also*
deforestation/logging
fossil fuels. *See* energy: fossil fuels as
source of
Freud, Sigmund, 181, 183
frogs, fungal disease of, 72–73. *See also*
species: extinction of

game theory, 83–107, 146, 160–161, 174, 187,
198, 200, 225; Prisoner's Dilemma, 83,
98; Public Goods Game, 96–98, 97;
Ultimatum Game, 96, 98
GAP (Aquatic GAP analysis program;
U.S. Geological Survey), 195–196
gasoline. *See* energy: gasoline/oil as
source of
Geographic Information Systems (GIS),
194, 196
glaciers, 41–42, 58
Global Footprint Network, 116, 213
global warming, 13, 40–48, 58, 73, 78, 104,
150, 158, 165, 181, 184, 186, 195
Gore, Al, 40, 170
government, 8, 40–41, 80–81, 86,
101–102, 104, 120, 135, 148, 154–155,
159–160, 165, 179–180, 182, 204, 208,
209, 211–213, 226; climate regulations of, 23; economic policies of,
156; environmental policies of, 108,
160, 169, 181, 189, 197, 202, 212, 228;
environmental regulations of, 87,
104–105, 139, 157, 164–166, 197, 200,
210; flood regulations of, 23;
policies of, 47, 109, 148, 176, 201;
population policies of, 16, 118, 145;
regulations of, 5, 85–87, 93, 101, 112,

118, 139, 143, 164–165, 176, 188, 210, 227
graphs, reading of, 12, 221, 223–224
Great Lakes, 65–66
green revolution, 150; and technology, 27
greenhouse effect, 3, 26, 34, 36–48, 54, 81, 102–103, 106, 153, 178–180; and runaway warming, 45–46. *See also* carbon dioxide
greenhouse gases 41, 104, 149, 165, 181
Griffith, Edgardo, 72
groundwater. *See* water: groundwater
Guam, 68–69, 75

habitats, 23, 29, 32, 35, 42, 45, 55, 57, 70, 76, 120, 133, 157–159, 171, 186; destruction (loss) of, 11, 36, 45, 54, 61, 63, 70, 75–76, 78, 104, 192–193, 209, 221; fragmentation of, 193; and invasive species, 64, 66, 69, 194; protection of, 104, 141, 210; restoration of, 168. *See also* species: extinction of
Hardin, Garrett, 84
Hawaiian Islands, 63, 69
Hayes, Tyrone, 59–60
human behavior, 9–10, 14, 30, 82–135, 160, 172, 190, 198, 208, 211; and advertising, 197–199; alteration of, 100, 144, 184, 197–198, 200, 205, 212, 226–227; altruism as, 83, 89–92, 94, 113, 116, 127, 130, 163, 198; attracting mates as, 123, 134; cheating as, 83–107, 110–111, 146, 183, 188, 200, 226–227; collective (social norms), 107, 173; control of, 107, 211; control of physical and biological surroundings (increasing survival) as, 116, 119, 133; cooperation (mutualism, reciprocity) as, 83–109, 117–118, 127–131, 135, 146, 155, 162–163, 183, 198, 226; as culturally determined, 172, 189; decision making as, 114, 139, 172, 197; denial as, 180, 183–184, 188; environmental effects of, 4–7, 109–110, 131, 152, 169, 195; and genetics, 89, 99, 127–131, 172; luxury consumption as, 26, 121, 126–127, 201; modeling of, 206, 213; modification of, 130, 146, 155, 227; moral obligation as, 99, 155; morality

as, 72, 75, 84, 90, 129, 133, 139, 153, 155–156, 161, 172–174, 202, 205, 207, 211, 228; prediction of, 30, 195; psychology, 109–110, 160–161, 198; punishment as, 83, 86, 89, 91, 95–96, 98–102, 105, 107, 111, 146, 165, 174, 199–200; rationalization as, 180–183, 188; and reproduction, 14–15, 108, 111, 114, 116, 117–119, 125–127, 133, 144, 159, 175–176, 187; selfishness as, 83–107, 116, 146, 155, 163–164, 173–174, 181, 187, 198, 207, 226; sociobiology on, 108–109; sublimation as, 183; suppression as, 183; and survival and comfort, 111, 118–119, 121, 127, 133, 144–145. *See also* evolution; game theory; religion; resources: human use (consumption, exploitation) of
human population, 4, 11–17, *13*, 26, 30, 32–33, 85, 88, 131–134, 138, 142–143, 149, 154, 160, 170–173, 188–189, 198; control of, 15, 35, 112, 118–119, 133, 141, 144–148, 176, 209–210, 213; growth of, 3, 8, 10–17, 26, 28–35, 63, 70, 84–85, 103, 109, 118, 135, 143–145, 150–151, 154, 164, 176, 186, 188, 193, 215–216; limits of, 14–15, 36. *See also* birth control
hurricanes. *See* weather: hurricanes
hydrologic cycle, 23, 25–26

incentives, 85–87, 89, 100–101, 112, 118, 127, 133–134, 136, 146, 160, 164, 177, 188, 199–200, 210
India, 1, 16, 42, 49, 57, 103, 116, 124, 145, 176–177
industrial revolution, 19, 36, 38, 87, 201
insecticides. *See* DDT; pesticides
intellectualization, 183
Intergovernmental Panel on Climate Change (IPCC; United Nations), 40–41
International Society for Ecological Economics, 159

Japan, 1, 7, 16, 116, 179, 185, 195

KiaOra, 6, 9, 97
Kingsolver, Barbara, 11
Kyoto Climate Summit, accord reached at, 149, 155, 181

Lake Victoria, 65
lakes, 3, 24, 59, 65, 67, 157; economic
   benefit of, 23; water clarity of, 157,
   203; water quality of, 45, 51, 53, 55, 80,
   158, 176. *See also* agriculture: fertilizer
   runoff from; pesticides; water;
   individual lakes
lesser developed countries. *See* developing countries
Lewis, Josiah, 6, 8, 86
lifestyle. *See* standard of living
Lomborg, Bjørn, 137
luxury fever, 121, 124

magic bullet, 147, 15, 191
malnutrition. *See* food: and malnutrition
Malthus, Thomas, 13
mammals, 128. *See also* species:
   extinction of; whales
Matthai, Wangari, 171
Maya, 15, 35, 121, 126–127, 154
McKibben, Bill, 3, 149
media, ix, 9, 105–107, 123, 137–138,
   208, 213
Middle East, 26, 43, 120
Millennium Ecosystem Assessment,
   5, 212
Millennium Survey (Gallup
   International), 8
Montreal Protocol, 49–50, 179
morals. *See* human behavior: morality as

National Academy of Sciences, 47
National Center for Environmental
   Economics, 159
National Ecological Observatory
   Network (NEON), 197
National Research Council, 86
Neumann, John von, 92
nitrogen, 4, 37, 51–52, 57, 93, 164, 219,
   220. *See also* agriculture: fertilizer
   use in
Northwest Passage, 1
nuclear war, 6, 92, 115, 178–181, 184–186
nutrients, 43, 150; concentrations of, 52,
   54, 76; cycling of, 23, 157; pollution by,
   52–55; in runoff, 157. *See also*
   agriculture; algae; nitrogen;
   phosphorus

oceans, 25, 37–38, 47, 54, 57, 73, 76, 156,
   197; coral (reefs) in, 45–46, 54, 75–76,
   93–94, 141; desalinization of, 25–26;
   evaporation from, 23, 25, 43; fisheries
   in, 3, 42, 75, 159, 204; flooding by,
   44–45, 180, 186; human effects on, 5,
   57; kelp in, 54, 73–75; level of, 41–45,
   47, 180, 186; marine ecosystems
   (habitats) in, 23, 35, 42, 47–48, 53–54,
   73–75, 158; pollution of, 3, 19, 54–55,
   57–58, 61, 69, 99; sediments in, 40, 43;
   temperature of, 41–46. *See also* water;
   individual oceans
Ogallala (High Plains) Aquifer, 24,
   27, 196
Orr, David, 106
overshoot, ix, 5, 14, 33, 115
ozone, 48–49; depletion of (ozone hole),
   3, 13, 34, 36–37, 48–49, *50*, 51, 64,
   80–81, 102, 104, 106, 137–138, 141, 150,
   153, 178–181, 187, 210, 219–220. *See also*
   CFCs

Payne, Roger, 58
pesticides, 3, 27, 43, 51, 55–61, *56*, 139, 141,
   143, 150; in runoff, 149
Pew Institute, 121
philosophy, 190, 205
phosphorus, 4, 51–52, 55, 219–220.
   See also agriculture: fertilizer use in
Pimentel, David, x
plants, 23, 35–40, 42–43, 45, 48,
   52–53, 56, 62–63, 66, 68–71, 74, 76,
   79, 93, 120, 133, 141, 150–151, 180,
   192, 194, 208. *See also* species:
   extinction of
plastic, 3, 49, 69, 151
policy. *See* government: policies of
politicians, 157, 164, 166–167, 178, 181, 197,
   199, 212
politics, 147, 175, 203, 209
polling/surveying (public opinion), 8,
   121, *122*, 154, 157, 181, 184, 197–200, 203,
   210, 226
pollution, 7, 23, 29, 31, 37–62, 66, 69, 80,
   130, 149–150, 153, 160, 167, 176–177, 183,
   189, 193, 196, 225–227; control of, 7,
   65–66, 87, 103–104, 120, 134, 141, 158,
   164, 166, 178, 188, 203–204, 209. *See*

also agriculture; atmosphere: pollution of; chemicals; nutrients; water: pollution of

poverty, 6–7, 32, 42, 99, 135, 146, 169, 178, 182, 201, 209, 212

precautionary principle, 139, 151

psychologists, 88, 100, 109, 110, 125, 197–200, 207

psychology. *See* human behavior

rabbits, 66, 68

regulation. *See* government: regulations of

religion, 89, 109, 114, 147, 153–156, 169, 174, 190, 205, 207

repression, 183

reproduction. *See* human behavior: and reproduction

reptiles 71. *See also* species: extinction of

reservoirs, 158, 196

resources, 3, 14, 100–103, 105, 107, 119, 133, 141, 161, 187, 189, 201, 207; aquatic, 21, 23–24, 26, 31, 143; conservation of, 86–87, 92, 95, 102, 121, 135–136, 138, 141–142, 159, 163, 168, 182, 210, 213; hoarding of, 124–127; human use (consumption, exploitation) of, 3, 5, 8–36, 63, 70, 83–93, 100–105, 107–112, 116, 119–121, 124, 127, 131, 133–147, 149, 154–156, 163–164, 169, 182, 185–188, 193, 202, 208–211, 213, 215; manage-ment (control) of, 105, 111, 123, 125, 127, 134, 136, 161, 164, 188, 211; value of, 84, 101, 136, 146, 156, 159, 178, 188, 202, 204. *See also* human behavior: attracting mates as; human behavior: luxury consumption as; sustainabil-ity; water

rivers, 23, 24, 26, 52, 65, 67, 158, 196; flooding by, 143; pollution of, 52, 55, 60, 158, 183. *See also* agriculture: fertilizer runoff from; nutrients; water

runoff, 26, 138, 143. *See also* agriculture; fertilizer runoff from; pesticides: in runoff; rivers

Russell, Bertrand, 83

Russia, 179

Schwartz, Steve, 91

science, 7, 8, 84, 106–107, 138, 147, 149–152, 173, 190–193, 206, 214

scientific journals, 137, 152–153

sea level. *See* oceans: level of

Simon, Julian, 13

social pressure, 15, 100, 112, 183

social science, 8, 98, 147, 160, 190, 196–197, 199

social scientists, 83, 85–86, 95, 109, 147, 161, 184, 196–202, 204, 206, 211, 228

Society for the Psychological Study of Social Issues, 161

sociobiology, 108–109

socioenvironmental restoration, 189–190, 198–199, 208

sociology, 152, 160

soil, 194, 208; carbon dioxide in, 38, 40, 46; erosion of, 31, 51, 66, 149, 171; and organism respiration, 46; tempera-ture of, 41, 43, 46

solar energy. *See* energy: solar

Southeast Asia, 32, 73

species, 107–108, 113, 119–120; abundance of, 63, 194; conservation of, 192–193; distribution of, 42–43, 63, 192–195; diversity of, 64, 68, 70, 76, 78; endangered, 29, 80, 93, 112, 193; extinction of, 10, 26, 34–36, 42, 45, 61–66, 70–79, 71, 85, 103–106, 131, 140–141, 151, 168, 187, 209, 211; inbreeding depression of, 192–193; interactions among, 93–94, 117, 128–129; introduced (exotic), 10, 36, 62–69, 81, 150, 209; invasive, 64–65, 69, 104, 141, 151, 186, 194, 196, 211; native, 63–65, 68–69, 79, 120, 191; rare, 66, 79, 192, 196, 210; unde-scribed, 71, 75–76; value of, 101, 105, 210. *See also* habitats: destruction (loss) of; human behavior: coopera-tion (mutualism, reciprocity) as; *individual species*

standard of living (lifestyle), 7, 11, 15, 17–18, 21, 29–32, 88, 103, 118–119, 124, 134, 142–144, 150, 152, 171, 177, 185, 188, 209, 211. *See also* developed countries; developing countries

statistical techniques, 192–194, 196, 199, 223
streams, 24, 27, 51–52, 72–73, 158–159, 184, 196; chemicals (pesticides) in, 55, 58–59, 167; pollution of, 167. *See also* agriculture: fertilizer runoff from; water
surveys, 8, 58, 59, 121, 122, 154, 157, 184, 198–199, 203, 226
sustainability, 86–89, 100, 134, 136, 138–140, 142, 144, 160, 164, 166, 169, 172, 175, 190–191, 201, 204, 206

Teal, Hannah, 2, 11, 90
technology, 10, 27, 29, 47–48, 85, 106, 110, 115–116, 120, 135, 142–144, 151, 175–176, 183, 185, 193, 195, 199, 201, 206, 225–228
temperature. *See* climate; global warming
tragedy of the commons, 84–91, 103, 135, 161, 187, 202
transcendence, 172, 212

ultraviolet (UV) radiation, 3, 36, 48, 50, 104, 138, 150, 178, 180
Union of Concerned Scientists, 149
United Nations: estimates of water shortages by, 21; estimates of world population by, 16
United States, 1, 7, 40, 43; biodiversity in, 64; ecological footprint of, 32; forest fires in, 45; introduced species in, 68–69; pollution in, 103, 116, 178, 181, 184, 187, 196, 213; population of, 4; religion in, 153, 155; resource use in, 88, 121, 134, 142, 148, 151, 168, 188, 208; standard of living in, 143, 213; water quality in, 26, 51, 53–55; water use in, 143
United States Society for Ecological Economics, 202–203
urbanization, 31, 32, 45, 51, 70, 104, 131–133, 132, 183

Veblen, Thorstein, 126
videophilic, 132
Vitousek, Peter, 4

Wallace, Alfred Russel, 40
water, 5, 21–27, 22, 32, 84, 120, 175, 187, 208; conservation of, 27, 99; drinking, 21, 24–26, 53, 157–158, 167, 175; in ecosystems, 27, 42, 73, 75, 79; evaporation of, 24–25; flow of, 193–194; groundwater, 24–25, 31, 55–56, 186; as habitats, 23, 53–54, 70, 159; pollution of, 21, 23, 24, 27, 36, 51–62, 80, 113, 167, 176–177; precipitation of, 23; quality of, 17, 24, 26, 29, 45, 51–52, 55, 157, 177, 182, 186, 193, 196; regulation of, 157–158, 196; supply of, 13, 21–27, 22, 29, 31, 34, 43, 80, 157–158, 161, 186; transfer (transport) of, 24–26; treatment of, 23, 53, 64–65, 157; use of, 17, 21–23, 22, 26, 27, 30, 141, 143–144, 149–150, 187, 189, 215–216; value of, 23; in watersheds, 23–24, 55, 158; in wetlands, 159. *See also* agriculture: irrigation in; hydrologic cycle; lakes; oceans; rivers; streams
weather, 25, 43, 165, 180; cyclones, 43–45, 44, 165; hurricanes, 43–45, 44, 96, 165, 186, 195; prediction of, 42. *See also* climate; global warming
wetlands. *See* water: in wetlands
Wetzel, Robert, 63
whales, 58, 63, 73, 74, 75, 101, 158
Wilson, Edward O., 108–109, 131, 189, 190
World Commission on Environment and Development, 173
World Food Program, 182
World Health Organization, 6, 44, 79

zebra mussels, 65–66, 67